Lenticular Imaging

Theory and Applications

A hundred years anniversary

Yitzhak Weissman

Published by Pop3DArt

An imprint of

Pop3DArt

P.O.B 12767, Herzliya 4673324, Israel
Email: **info@pop3dart.com**
Web: **www.pop3dart.com**
Book web-page: **www.pop3dart.com/lenticular-imaging**

Revision 1.02, January 2018

Editions

Description	**Color**	**ISBN**
Hardbound print	Full	9789655724035
Softbound print	Black and White	9789655724912
e-book	Full	9789655724028

Dedication

This book is dedicated to the pioneers of lenticular imaging, who, with genius and devotion started all this a hundred years ago.

Preface

One day, about two years ago, I woke up sweating with a pain in my chest. Panicking, I reached out for the home ECG recorder and attached its electrodes to my trembling body. A moment later, the signal was transmitted to my doctor.

It turned out to be a false alarm: my heart was healthy and sound. But when I went to my office later, the first thing I did was to write down an inventory of knowledge which resided in my head without any documentation. Staring at the list, I realized the sheer size of my research into lenticular imaging over the past five years. With a fresh memory of the morning's incident in my mind, I decided to write this book.

When I started the project, I knew that additional research would be necessary, but I surmised that this would be limited to filling up a few gaps. Well, I was wrong. As the work progressed, the "gaps" widened and unveiled wide fields of the unknown. At times, I felt like an explorer traveling in an unchartered land abundant with hidden treasures. Extensive research interleaved with the writing took two years to complete.

The work on my earlier book ("Optical Network Theory") during the fateful 90/91 winter was interrupted repeatedly by ballistic missile raids. Fortunately, in the past two years fortune favored me. Nothing tragic or dramatic happened in my personal life, and the most exciting event was the birth of my beloved grandson Aviv a year ago. I could work and complete the book with a peaceful mind.

Finally, I have allowed myself to add a personal touch by including a citation following every chapter title. These citations are not related to the content of their corresponding chapters; but they reflect ideas which inspired me in recent years. They serve as a light diversion form the strict scientific text, and, maybe, they will inspire you too.

Yitzhak Weissman

Herzliya, Israel, August 2017

Acknowledgements and credits

In writing the manuscript, I had the great privilege of obtaining generous help from several outstanding individuals. Stephen Palmer, Barry Johnson, Gary Jacobsen and Tomer Czyzewski reviewed early versions of some chapters and provided valuable feedback. Michal Husak and Paul Bourke reviewed the first version of the complete manuscript. Paul's enlightening comments caused a major revision, and thanks to him the manuscript is much improved.

The manuscript was written, designed and formatted in Microsoft Word 2016. I found its fancy equation editor unusable, so I used Microsoft Equation 3 instead to enter the mathematical expressions. Most figures were done with Visio 2013. Others were made with Photoshop and DAZ Studio. The optical resolution study was done with Matlab.

Lisa ("WordDoctor" from Servicescape) proofread an early version of the manuscript meticulously, corrected all typo and grammar mistakes, and polished my style. I used Grammarly to proofread the final text.

Anna Castle is not only an accomplished mystery book author, but also a skillful Adobe Acrobat expert. Her script (Castle, 2015), made generously available to the public, saved me a lot of time and agony in converting the manuscript to a file suitable for submission to print.

Contents

Chapter 1
Introduction

"For there was a moment when anything was possible. And there will be a moment when nothing is possible. But in between we can create."

- Moshin Hamid, in "How to Get Filthy Rich in Rising Asia"

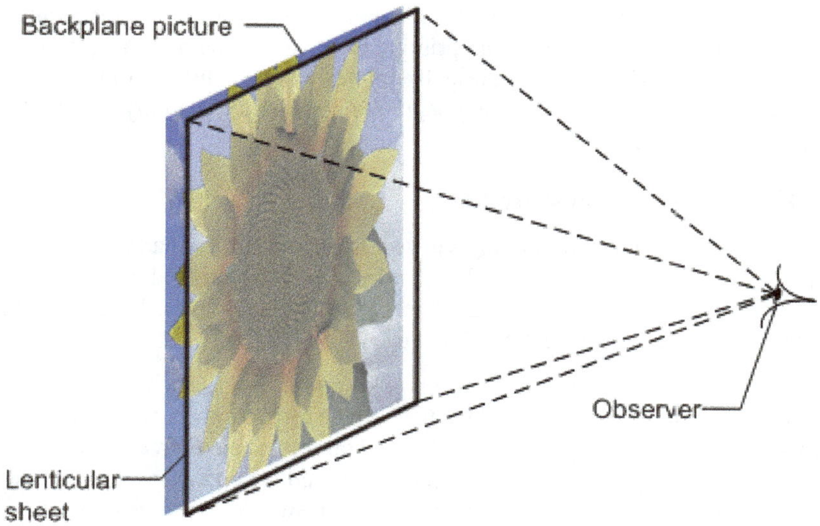

1.1 About this book

This book is intended for students, scientists, and engineers wishing to learn about lenticular imaging or to expand and diversify their knowledge about 3D display systems in general. It will also be useful to practitioners of lenticular print technology, 3D photographers, and graphical artists who wish to understand and explore the lenticular medium for imaging.

The book treats phenomena and theory related to lenticular imaging with backplane pictures, giving the reader a solid background on which they can base their exploration or development. The priority in the book is to explain the theoretical principles, and in most cases, detailed examples supplement the academic discussion. Nevertheless, this book is not a practical guide for the fabrication of lenticular pictures.

There are two main types of lenticular pictures: 3D and dynamic. A 3D picture displays a three-dimensional scene by invoking stereopsis (Stereopsis, 2017). Dynamic pictures display different and generally independent images for different viewing angles. Formally, 3D pictures are a particular case of dynamic pictures, but they will be regarded in this book as a separate category, adopting the terminology in the art. The focus in this book is on 3D pictures, but the discussion of general dynamic pictures is also included.

This book exhibits original and hitherto unpublished research. The disciplines involved are optics, image processing, printing technology, computer graphics, and photography. Basic background in these disciplines will be helpful for the reading. The mathematical treatment is with varying degree of complexity, but only college-level mathematical tools are used.

1.2 History and background

The first technique that allowed freely-viewed 3D pictures to be made was the "parallax barrier" (Parallax barrier, 2017), and is over 100 years old. It was independently invented by Berthier (Berthier, 1896) and Ives (Ives, 1902), and was later patented by Ives (US Patent No. 725,567, 1903).

The parallax barrier method is based on obscuration, and as such, it suffers from low optical efficiency. Besides, the displayed image includes the grid of the obscuring elements. The lenticular method, which is based on refraction, is free from these shortcomings. Despite this, the parallax barrier method is popular in stereoscopic electronic displays, because it allows easy electronic switching between 2D and 3D modes.

Lenticular imaging is a one-dimensional implementation of the integral imaging concept (Lippman, 1908). In contrast to true integral imaging that uses a dense array of plano-convex round lenses and can display a change of perspective in both the horizontal and the vertical directions, lenticular imaging uses an array of plano-convex cylindrical lenses instead. With such an array, it is possible to show the changing perspective in one direction only, but this is sufficient to create an impressive 3D effect.

The first description of a 3D lenticular picture is due to Hess (US Patent No. 1,128,979, 1915), who proposed using a lenticular sheet to create a stereoscopic picture. Kanolt generalized this idea and showed how to make a multi-view lenticular picture (US Patent No. 1,260,682, 1918). Kanolt's scheme forms the basis for modern lenticular imaging.

The development of lenticular imaging started in the late 1920s by Herbert Ives and Harvey Prever in the US and a few other scientists (Vance, 2000). Soon after that, the unique imaging capabilities of a lenticular sheets were established, and researchers in Britain, France, and Germany developed processes to manufacture such sheets.

In the early 1940's Bonnet in France and Winnek in the US developed photographic processes and cameras to record images suitable for 3D lenticular imaging. Bonnet opened a studio on the Champs-Elysees (Paris, 1942), which included a 3D lenticular photography service.

The development of mass-production techniques for lenticular sheets in the early 1960s made lenticular imaging available to advertising and publishing markets. *Look* magazine published the first mass-produced color lenticular picture in 1964, using a process developed by Kodak. Over 8 million copies were sold.

Early methods for image deposition on a lenticular sheet were based on photography, as proposed originally by Kanolt (US Patent No. 1,260,682, 1918). For such use, the lens sheet was coated with a photographic emulsion. Starting in the 1980's, rapid developments in the print industry shifted the interest to image deposition using lithographic printing. This development brought significant cost reduction, and the use of lenticular pictures, both dynamic and 3D, became prolific in advertising, packaging, and novelty item industries.

Lithographic printing is cost-effective only at high volume, typically at several thousands of units, and is unsuitable for a thick lenticular sheet. The advent of high-resolution digital printing and digital image processing beginning around 1990, allowed the creation of a single lenticular picture at an affordable price and any sheet thickness. This price drop opened the possibility of making single pictures for personal use and art at a reasonable cost.

Concerning scientific research, lenticular imaging stayed in darkness for many decades. Perhaps the last and the only significant academic study of lenticular imaging is from 1976 (Okoshi, 1976). This has changed in the past 20 years, during which considerable efforts had been invested in the development of electronic lenticular displays. In such systems, an LCD is used to create the backplane picture, and therefore they can display a 3D video which can be watched without any visual aids. This technology attracted much attention from commercial display companies, in the hope of using it for consumer TV sets. The business activity spurred scholarly research (Leuder, 2012) (Byoungho, Soon-gi, Keehoon, & Jisoo, 2016), which also included some aspects of lenticular imaging. Although electronic and printed lenticular pictures share a few essential elements, they also differ in several important aspects, the most important being their pixel structure. The particular pixel structure of LCD displays requires slanting of the lenticular sheet.

A special chapter in this book is devoted to electronic lenticular displays which use slanted lenticular sheets. To simplify the exposition, in all other chapters the theory is presented for the simpler case of non-slanted ("erect") pictures, and the backplane picture is assumed to be printed. This special chapter outlines the application of non-slanted displays theory to slanted displays. However, the rigorous treatment of imaging in slanted displays requires further research.

1.3 The grids of lenticular imaging

The lenticular picture is composed of two basic elements: the lenticular sheet and the backplane picture. Each one of these elements is characterized by one or more grids. The primary technical challenge in the fabrication of the lenticular picture is the management, adjustment, and alignment of these grids.

In modern lenticular imaging, the backplane picture is prepared by image processing and printing. In such a workflow, there are three main grids:

1. The lenticular array grid,
2. The optical grid,
3. The printed grid.

Each one of these grids has a characteristic pitch. These quantities are frequently mentioned throughout the book and are denoted by the same symbols in the different chapters.

Table 1: The three main grids of the lenticular picture

Grid name	Pitch symbol
Lenticular	p
Optical	p_o
Printed	p_p

The optical grid is manifested by an array of bright lines which appear at the lenticular sheet backplane when illuminated from the front by a point light source. In the optical sense, these lines are the images of the light source as created by the lenticules of the lenticular sheet. The position of this grid depends on the location of the light source, and its pitch depends on the distance of the light source from the lenticular sheet. When the light source is turned off, the optical grid disappears. Throughout this book, the optical grid is an imaginary grid, which would have appeared if a point light source was positioned in the center of the observer's eye pupil. By contrast, both the printed and the lenticular grids are physical and permanent entities.

The lenticular and the printed grid pitches are entirely determined by the lenticular sheet and the printing process respectively. The optical grid pitch is determined by the lenticular sheet characteristics and the viewing distance. In the early lenticular pictures, the backplane picture was created photographically and was a developed photographic emulsion. In this method, the "printed" and the optical grids pitches coincide.

The primary task in the printing process is matching the optical and the printed grids for a given viewing distance. The main task in the picture fabrication is the alignment of the printed and the optical grids so that they are centered and parallel to each other.

Digital technology has revolutionized all industries, by boosting performance and dropping prices. However, this technique requires quantization of the relevant physical quantities so that they can be represented by numbers with a limited number of bits. In the context of digital imaging, the quantization requirement gave rise to the pixel grid. This is an additional grid which is present in lenticular imaging. Even though this grid was not included in the list of fundamental grids, its existence has an important impact on lenticular pictures, particularly in cases in which its pitch nearly matches the printed pitch. The issues related to the pixel grid are discussed in paragraph 3.4.

Finally, there is yet another grid which affects lenticular pictures, and which is purely a printing technology feature. This grid is called “screen” and is used for the creation of color gradients from discrete dots (Halftone, 2017). The existence of this grid limits the pixel densities used to print the backplane image to a series of values called “preferred densities." This fact is discussed and illustrated in paragraph 10.4.

1.4 Overview

The chapters of this book are arranged in a logical order. Namely, each chapter depends on concepts introduced and discussed in the previous ones. However, with little imagination and intuition, they can be read independently. All chapters rely on the grid concepts, and therefore the reader should understand their significance as explained in the earlier paragraph, before trying to learn a chosen chapter independently. It is also recommended for all readers to read Chapter 2 before attempting to read another chapter independently.

There are four main tasks in making a lenticular picture:

1. Content preparation,
2. Calibration,
3. Interlacing,
4. Fabrication.

Fortunately, these tasks are to a large extent independent, and each can be accomplished by different personnel. Usually, the first step is carried out by a graphics artist, and the rest by the printing studio operators.

The outcome of the content preparation is a sequence of images, called a “lenticular sequence.” These images are multiplexed into a single “interlaced” image. The interlaced image is printed and (if not printed directly on) bonded to the lenticular sheet. Calibration is necessary for the printing. All principles upon which these tasks are based are described in this book. Other topics presented in the book are related to the visual characteristics and performance of lenticular pictures.

The basis for understanding the lenticular picture mechanism is the optics of the lenticular sheet. This topic is the subject of Chapter 2. Many concepts introduced in this chapter are borrowed from the discipline of classical imaging optics, but many others are unique to cylindrical lens and lenticular sheets. These ideas are used in many instances throughout the book.

Chapter 3 includes several topics related to image processing, which are necessary for the discussion in subsequent chapters. Most of the issues are well known in the image processing discipline and are reproduced here for the convenience of the reader. There are others which are less familiar, like the matched pixel grid. This concept is important for the understanding of the visual characteristics of lenticular pictures.

The interlacing process is discussed in detail in Chapter 4. This discussion is perhaps the most abstract exposition in the book. The general interlacing algorithm may be implemented in many variations, the study of which is full of intricate technical details. This study is not included in the present book.

Calibration is the measurement of the optical grid pitch. This measurement must be done with a high accuracy, typically with a relative tolerance of the order of 10^{-4}. Variations on this scale exist in the manufacturing of a lenticular sheet, as well as between different rulers (like the ruler of the lenticular sheet extrusion machine and the printer of the printing studio). Therefore, it is necessary to perform the measurement in the printing studio, preferably for each lenticular sheet. This process is the subject of Chapter 5.

Practitioners working with the 3D lenticular medium know that objects imaged near the picture plane are displayed with a high visual quality. As the distance between an imaged object and the picture plane increases, its display starts to acquire visual defects, and its visual quality deteriorates. As the most valuable visual asset of a 3D lenticular picture is the display of depth, artists strive to display as much depth as possible. Consequently, practitioners often choose to allow defects with a certain magnitude to increase the displayed depth. Quite often, the optimal compromise between depth display and visual quality is achieved by an expensive and time-consuming iterative process of try and correct.

The matter is complicated by the fact that there are two different mechanisms, each one responsible for a different visual defect. It is possible to trade one defect for the other, but not to reduce both simultaneously. In Chapter 6 it is explained how to control the compromise between depth display and visual defects to optimize the picture display without the need to perform expensive try-and-correct experiments.

Chapter 7 is a review of electronic lenticular displays. The main differences between printed and electronic lenticular displays are discussed, and the methods which are used to create electronic displays are outlined, using the concepts of printed pictures.

Readers who have experience viewing 3D lenticular pictures know that there are viewing regions in which the picture looks crisp and clear and regions in which the picture exhibits blurriness and even noticeable visual defects. Understanding this phenomenon is essential for the picture design, and this is the topic of Chapter 8.

There are two methods to create a lenticular sequence for a 3D picture: synthetic and photographic. In the synthetic method, the sequence is created from a set of two-dimensional images using various algorithms. A typical procedure is to use several images disposed on layers with different depths. The layers can be further sculptured using depth maps (Depth map, 2017). In Chapter 9, the photographic method is discussed in detail. The discussion applies to both real and virtual photography, which is used for scenes generated by a 3D software program. This chapter is illustrated with numerous examples, and is closer to be a practical guide than the other chapters.

The characteristics of lenticular pictures depend heavily on the characteristics of the printing process. Some aspects of the digital print which are relevant to lenticular imaging are discussed in Chapter 10. However, this subject requires further research.

Chapter 11 is devoted to the analysis of the optical resolution and distortion of a lenticular sheet. These mechanisms set the most fundamental limit on the performance of lenticular images. For digitally printed and electronic pictures, the bottleneck is the picture resolution. Higher-resolution printing processes may compete with the optical resolution. The optical resolution analysis explains a few phenomena, for instance, the blurriness that appears when viewing the picture at high horizontal angles.

Chapter 2
Basic optics of lenticules and lenticular sheets

"Not only could he simplify complex matters, he could also keep simple matters simple ."

- David McCullough, in "Truman"

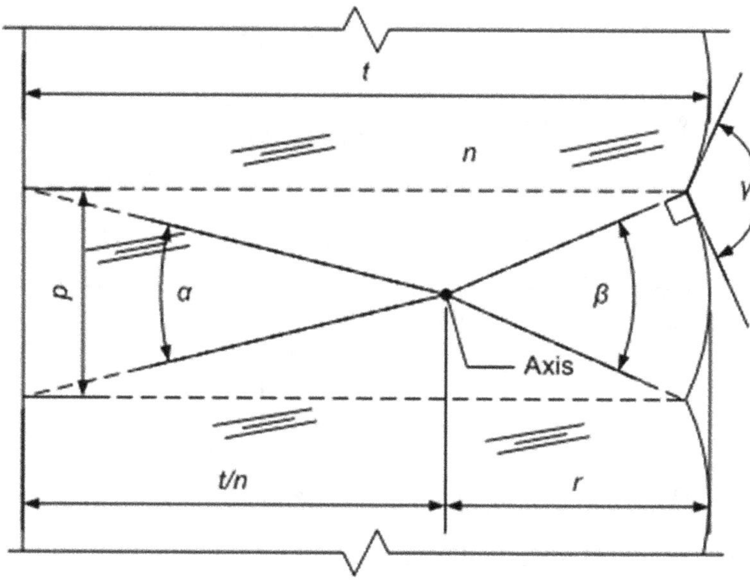

2.1 Cylindrical lens geometry

A lenticular sheet is an array of identical lenticules. Each lenticule is an optical element known as "cylindrical lens," of the type called "plano-convex."

The geometry of a general plano-convex cylindrical lens is shown in Figure 1. It may be regarded as a union of two parts: a relatively thin cylindrical section and a solid cuboid. The cylindrical section is responsible for the imaging function of the lens. The lens geometry is characterized by its dimensions and the radius of the parent cylinder. This radius is also called the "radius of curvature" of the lens.

The parent cylinder axis will be referred to as the lens axis. Unlike the axis of circular plano-convex lenses, customarily defined as orthogonal to its planar face, this axis is parallel to the lens backplane.

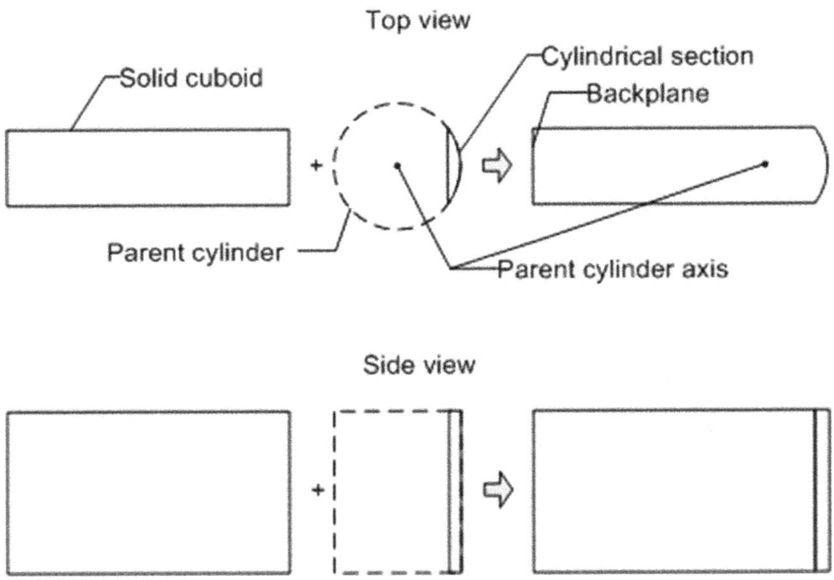

Figure 1: Plano-convex cylindrical lens geometry.

2.2 Fundamental optical characteristics

Let us consider a cross-section perpendicular to the lens axis shown in Figure 2. The curved surface in this cross-section is a circular arc. Such an element is known to exhibit optical imaging. This means that all rays (in this plane) emerging from a point O in front of the curved surface will approximately converge to a point F behind it. This is illustrated in Figure 2 where three distances were introduced: s', s, and f. These distances denote the source and the convergence point distances from the curved surface (s' and s respectively) and the associated focal length (f). The fourth distance t in Figure 2 denotes the lens thickness.

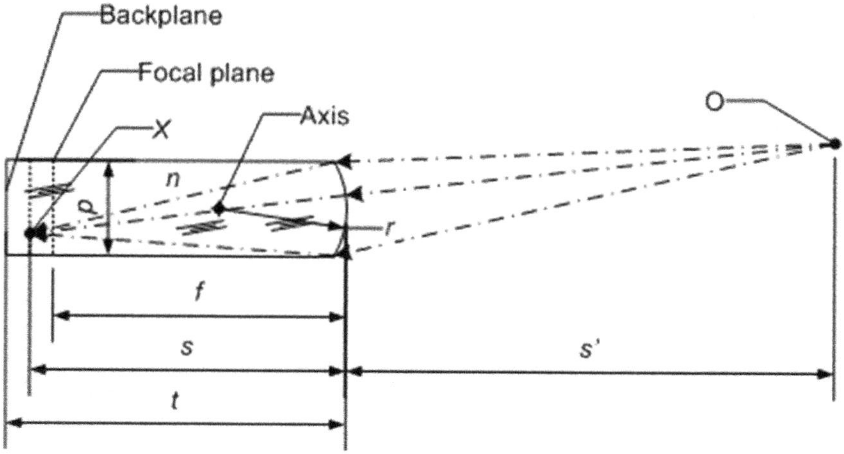

Figure 2: Fundamental optical characteristics of a lenticule.

The distances *s'*, *s*, and *f* are related by the known "lens maker's formula" (Jenkins & White, 1957):

$$\frac{n}{s}+\frac{1}{s'}=\frac{n}{f}, \qquad \textit{Equation 1}$$

where *n* is the lens material index of refraction. The plane located at distance *f* behind the curved surface is called the "focal plane," and its footprint is shown in Figure 2. Rays emerging from infinity will converge to points on this plane, as can be derived from Equation 1 by letting *s'* increase to infinity.

The focal length *f* is related to the lens radius *r* by (Jenkins & White, 1957):

$$\frac{n}{f}=\frac{n-1}{r}. \qquad \textit{Equation 2}$$

Lenticules are designed so that their backplane coincides with their focal plane, or, in other words, $t = f$. Substituting *t* for *f* in Equation 2 results in

$$r=\frac{(n-1)t}{n}, \qquad \textit{Equation 3}$$

or

$$t=\frac{nr}{n-1}. \qquad \textit{Equation 4}$$

In general, the parameters *t*, *r*, and *p* are independent. The requirement that the focal plane will coincide with the backplane of the lens reduces the number of free parameters that define a lenticule to just two. The geometry of a lenticule with the focal plane at the backplane is shown in Figure 3.

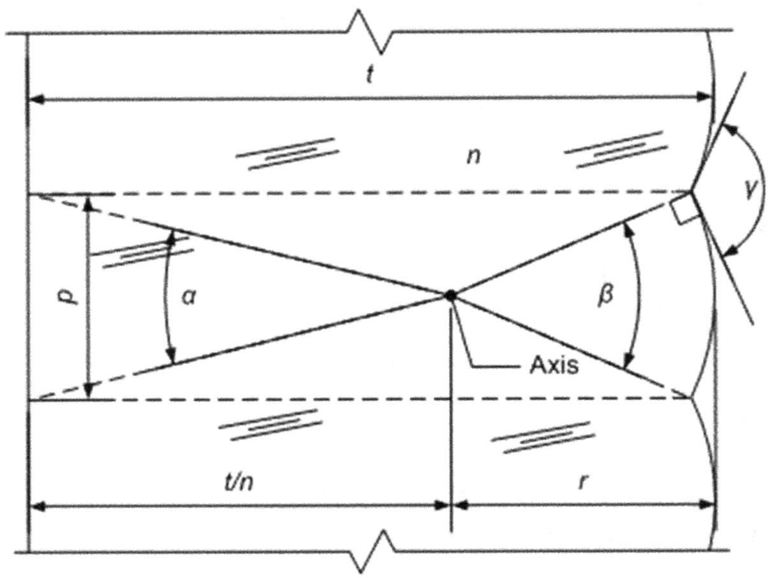

Figure 3: The geometry of a lenticule.

Transparent·optical polymers are used to manufacture lenticular sheets. The indices of refraction of such materials typically vary between 1.49 to 1.6. Some values are given in Table 2. In all calculations presented in this book, it will be assumed that the value of n is 1.6.

Table 2: Indices of refraction of optically transparent polymers used for lenticular lenses.

Material	**Index of refraction**
PVC	1.60 (Scientific Polymer, 2015)
PETg	1.57 (List of refractive indices, 2015)
Polystyrene	1.5894 (List of refractive indices, 2015)
Poly-carbonate	1.5848 (Refractive Index Info, 2015)
Acryl	1.490 – 1.492 (List of refractive indices, 2015)
PMMA	1.4893 – 1.4899 (List of refractive indices, 2015)

In the theory of optical imaging lenses, there is an important parameter called "relative aperture" (also known as "f-number," "focal ratio," "f-stop," etc.). This parameter is the ratio between the lens' focal length f and the diameter D of its entrance pupil (Jenkins & White, 1957):

$$R_a = \frac{f}{D} \,. \qquad \textit{Equation 5}$$

For lenticules, a similar parameter is introduced: relative pitch. It will be denoted by R_p and defined as

$$R_p = \frac{np}{t} \,. \qquad \textit{Equation 6}$$

Since in lenticules the thickness t is the same as the focal length f, and p plays the role of the aperture size, R_a is related to the inverse of R_p. In the present treatment, the definition in Equation 6 is more convenient than the traditional definition used for R_a.

There are two primary applications of lenticular pictures: 3D and dynamic. The value of R_p in lenticules used for 3D applications is in the range 0.4 to 0.6. In dynamic pictures, values of R_p between 1 and 1.3 are typically used.

Figure 3 shows three angles which play an essential role in lenticule optics: the lenticule viewing angle α, the lenticule acceptance angle β, and the lenticule vignetting angle γ.

The viewing angle is defined as the angle subtended by the backplane with respect to the lenticule axis, as shown in Figure 3. It is seen that

$$\tan\left(\frac{\alpha}{2}\right) = \frac{R_p}{2} \qquad \textit{Equation 7}$$

For small lenticule viewing angles, one can approximate (in radians)

$$\alpha \approx R_p \,. \qquad \textit{Equation 8}$$

The lenticule acceptance angle is given by

$$\tan\left(\frac{\beta}{2}\right) = \frac{p}{2r} = \frac{R_p}{2(n-1)} \,. \qquad \textit{Equation 9}$$

For small lenticule acceptance angles one can approximate (in radians)

$$\beta \approx \frac{R_p}{n-1} \,. \qquad \textit{Equation 10}$$

For values of n shown in Table 2, the angle β is larger than α. Both the viewing and the acceptance angles characterize the imaging properties of the lenticule and the lenticular sheet.

The vignetting angle γ is the angle between the two tangents to the lenticules surfaces at the crevice between them. It affects viewing at a large angle, as illustrated in Figure 4. When the lenticular sheet is observed from an angle which exceeds $\gamma/2$, the crevice line between adjacent lenticules becomes obscured, and the observer will not be able to see their full aperture.

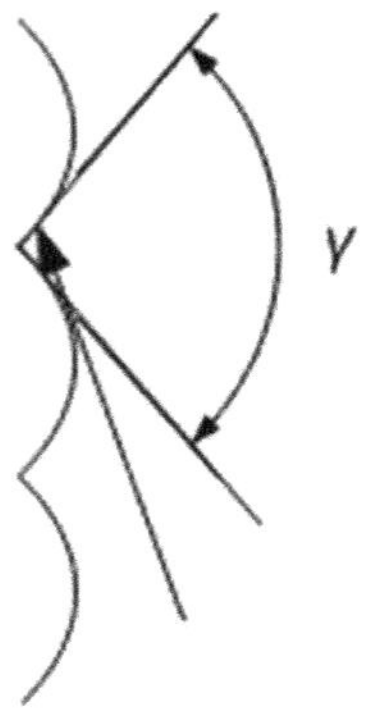

Figure 4: Obscuration caused by viewing beyond the vignetting angle.

Since the tangent to a circle is orthogonal to its corresponding radius vector, it can be deduced (with the help of Figure 3) that (in radians)

$$\gamma = \pi - \beta \,.$$

Thus, the vignetting angle is smaller for lenticules with a higher value of R_p.

Table 3 shows values of the various quantities discussed here in lenticules used by commercial manufacturers. In the first two columns, the most common specifications of lenticular sheets were entered: the thickness t and the lenticule density ρ, which is the number of lenticules per unit length. It is common to use an inch as the unit of length for density, in which case the density unit is called "lpi," standing for "lenticules per inch." The lenticule pitch p can be derived from the density by the relation

$$p(\text{mm}) = \frac{25.4}{\rho(\text{lpi})} \,.$$

The values in the t, p, and r columns are in units of mm. The values in the lenticule density column are in units of 1/inch. All angular values are in degrees. The first two columns (shown in bold types) were specified, and all other columns were calculated with $n = 1.6$.

Table 3: Characteristics of some common lenticular sheets

t (mm)	ρ (1/inch)	p (mm)	r (mm)	R_p	α	β	γ
0.3	**100**	0.25	0.11	1.35	68°	97°	83°
1	**70**	0.36	0.38	0.58	32°	52°	175°
1	**40**	0.64	0.38	1.02	54°	81°	99°
1.2	**60**	0.42	0.45	0.56	32°	50°	130°
2	**40**	0.64	0.75	0.51	29°	46°	134°
4	**25**	1.02	1.50	0.41	23°	37°	143°
4	**20**	1.27	1.50	0.51	29°	46°	134°
6	**16**	1.59	2.25	0.42	24°	39°	141°

2.3 Basic concepts for the description of lenticule optics

Two families of planes are of interest in relation to the lenticule geometry:

- Axial planes: all planes which contain the lens axis, and cross the lenticule curved surface,
- Imaging planes: all planes perpendicular to the lens axis.

Both families are infinite, and any axial plane is perpendicular to any imaging plane. Figure 5 shows representatives of both families. The intersection contours of these planes with the lens surface are highlighted in Figure 5 with broken lines.

The imaging plane containing the observation point will be called “viewing imaging plane." If the viewing direction is within the acceptance angle interval, then there is also an axial plane containing the observation point, and it will be called correspondingly "viewing axial plane." For viewing directions outside the acceptance angle interval, there are no viewing axial planes.

Light rays (Jenkins & White, 1957) are a useful concept to understand and analyze optical systems. This concept will be used in this chapter and the analysis of optical resolution (Chapter 11). The imaging and the axial viewing planes are used to define two unique ray families:

- Imaging rays: rays emitted by the observer’s pupil center and contained in the viewing imaging plane,
- Axial rays: rays emitted by the observer’s pupil center and contained in the viewing axial plane.

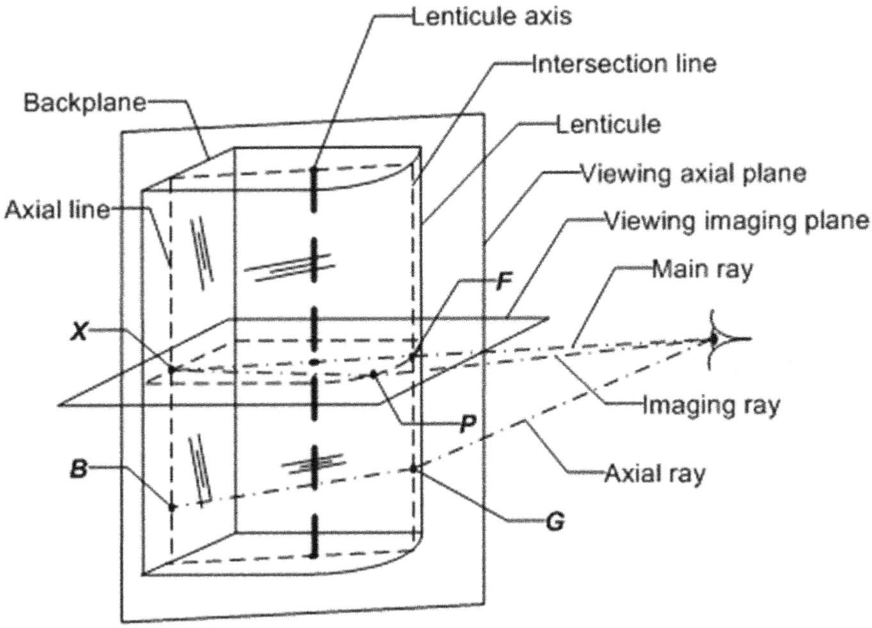

Figure 5: Fundamental concepts in the optical analysis of a lenticule.

For any given observation point within the acceptance angle, there is one ray called the "main ray" which belongs to both families. The main ray is not refracted by the curved surface because it is incident in the normal direction. Representatives from each ray family, as well as the main ray, are shown in Figure 5.

For a given observation point, all imaging ray trajectories are contained in the same plane. The same is true for all axial rays. Therefore, these rays can be analyzed in two dimensions, which is much simpler than an analysis in three dimensions. All rays emitted by the observer's pupil center which are neither axial nor imaging are called "skew rays."

2.4 The visible zone of a lenticule

An observer can see only points which emit rays that enter the eye's pupil. Such points are called "visible points." The set of all visible points on the lenticule backplane forms the visible zone.

According to the laws of classical optics, all imaging rays emerging from the center of the observer's eye converge at a certain point on the lenticule backplane. In Figure 5, the convergence point is denoted by X and is located on the axial line. Since the ray trajectories are reciprocal, the opposite is also true: all imaging rays emerging from point X (and refracted by the curved surface) will converge at the observer's eye. Therefore, point X is visible.

All axial rays stay after the refraction in the viewing axial plane and therefore intersect the backplane on the axial line. The reader can easily convince themselves that for any point on the axial line there is a corresponding axial ray. Therefore, the whole axial line is visible, not only point *X*.

Quantitative derivation of the visible zone geometry requires 3D ray tracing of skew rays. This analysis (Chapter 11) shows that, to a good approximation, the visible zone of a single lenticule is a thin line of finite thickness with the axial line passing near its center. The approximation in which the visible zone is a thin column with the axial line passing through its center will be called "simplified model." The finite width of the visible zone is due to optical aberrations and other mechanisms and is usually much smaller than the lenticule width.

2.5 The visible zone of a lenticular sheet: the optical grid

A lenticular sheet is an array of identical and parallel lenticules with a common backplane. The visible zone of the lenticular sheet consists of the visible zones of the individual lenticules, which form an array of equidistant and parallel lines. This visible zone of the lenticular sheet will be called "optical grid." An observer watching a picture covered by a lenticular sheet will see only the portions which overlap the optical grid. These portions are a tiny fraction of the picture area.

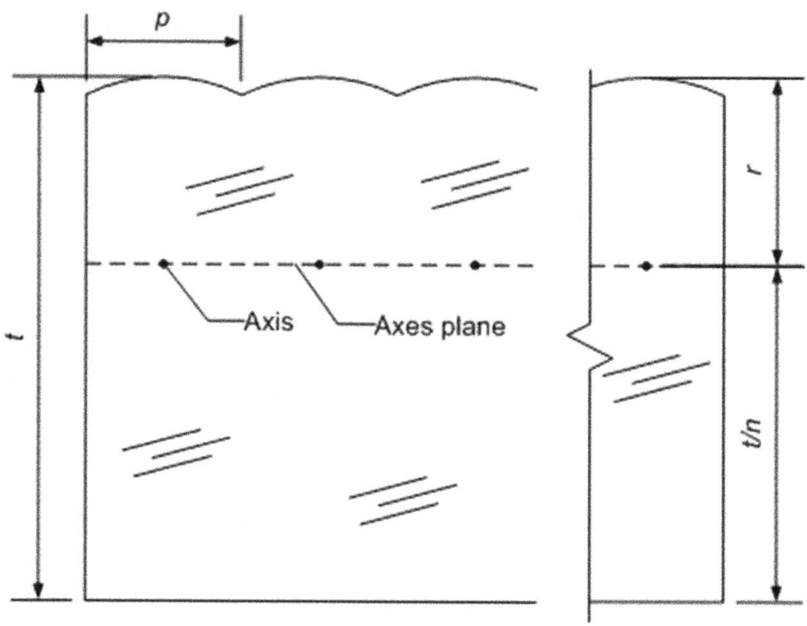

Figure 6: Cross-section of a lenticular sheet.

The position of the optical grid depends on the position of the observer's eye. When the observer moves, the optical grid moves in general too.

Figure 7 shows an imaging plane cross-section of a lenticular sheet with an observation point A.

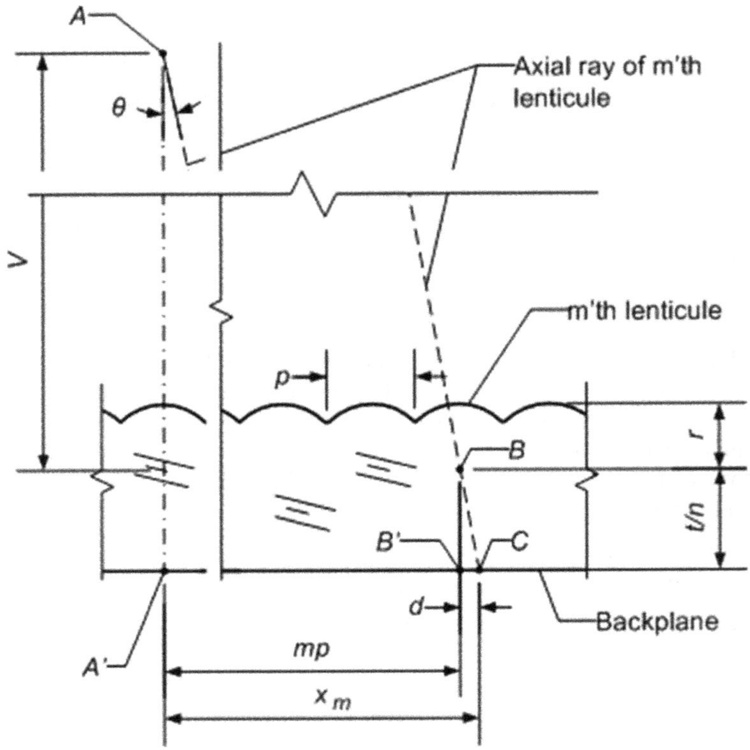

Figure 7: Diagram for the calculation of the axial grid pitch.

For the sake of simplicity, it will be assumed that the normal from point A to the sheet backplane passes through the axis of one of the lenticules, and crosses the backplane at point A'. The m'th lenticule axis is denoted by B, and the intersection of the m'th axial ray with the backplane by C. Let x_m be the distance of C from A'. From Figure 7 it follows that

$$x_m = mp + d.$$

Using triangular similarity relation for triangles $A'AC$ and $B'BC$ one gets

$$d = \frac{mp}{V}\frac{t}{n},$$

so that

$$x_m = mp\left(1 + \frac{t}{nV}\right). \qquad \text{Equation 11}$$

The distance p_a between two adjacent axial lines is

$$p_a = x_{m+1} - x_m = p\left(1+\frac{t}{nV}\right). \qquad \textit{Equation 12}$$

According to this relation, p_a does not depend on m, which means that the axial grid pitch is constant. Therefore, the axial lines form a regular grid, called "axial grid." Its pitch is different from the pitch p of the lenticular sheet and depends on the observation distance V.

Equation 12 plays a fundamental role in lenticular imaging. For the convenience of the reader, the various parameters which appear in it with their description are listed in Table 4.

Table 4: The parameters of Equation 12.

Symbol	Description
p_a	Axial grid pitch
p	Lenticular sheet pitch
t	Lenticular sheet thickness
n	Lenticular sheet material index of refraction
V	Viewing distance

Within the simplified model, the optical gridlines centers coincide with the corresponding axial lines. Therefore, according to this model p_a is also the pitch p_o of the optical grid.

Note that this derivation breaks down when the angle θ exceeds half of the acceptance angle. In this case, the ray originating from the observation point A and directed to the axis of a given lenticule will cross the lenticular sheet at the curved surface of a neighbor lenticule and will be refracted. In such circumstances, there is no axial ray, so also the concept of the simplified model breaks down. To understand what happens in this case, one must resort to a ray-tracing analysis, as presented in Chapter 11. This analysis shows that outside the acceptance angle interval, the spacing between the optical gridlines becomes irregular, and their widths significantly increase with θ.

2.6 Optical imaging with an observer at the image plane

The optical effect of a lenticular sheet on a picture is considerably different from familiar optical imaging effects, and its explanation is complicated due to its three-dimensional nature. Some features of this effect can be understood with concepts of classical optical imaging. Let us consider an object and its optical image created by a thin imaging lens as shown in Figure 8. Let the observer be located at a convenient viewing distance from the optical image (more than 30cm), so their eye can focus on it. In this case, the observer will see a magnified and inverted image of the object.

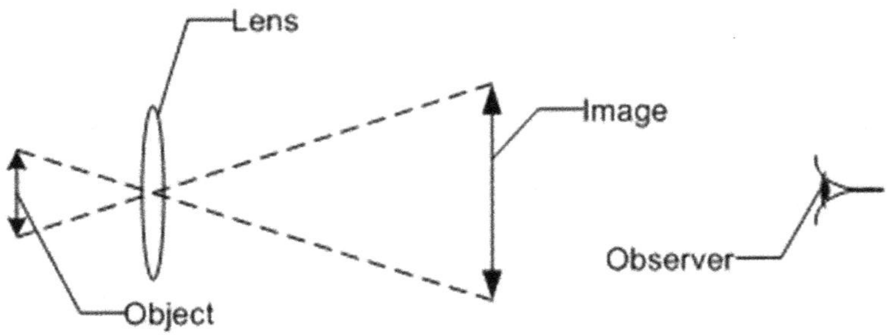

Figure 8: Common optical imaging.

Figure 9 shows an uncommon case, in which the eye of the observer is in the optical image plane. In this case, the observer will not see the optical image. Instead, they will see the graphical content from a small disk at the center of the object, shown as the "visible zone." The diameter of this disk is determined by several factors, like the eye's pupil finite size and the lens aberrations. This will be discussed in more detail in Chapter 11.

The eye of the observer cannot focus on the image and will focus instead on the lens (assuming again that the lens is at a viewing distance from the eye). The observer will see the viewing zone magnified to the lens aperture, but any graphical content contained in the viewing zone will be scrambled. The lens aperture will appear to have an approximately uniform color, which will represent a kind of average of the colors contained in the viewing zone.

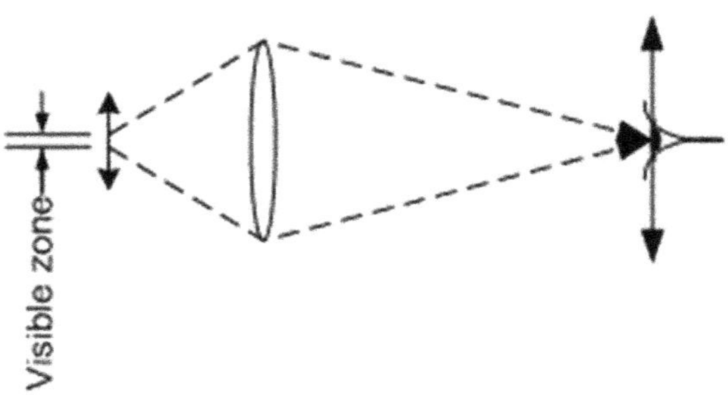

Figure 9: Imaging with the observer at the image plane.

2.7 Viewing geometry of a lenticular picture

The concepts needed to discuss the visual effects of a lenticular picture are shown in Figure 10. It is always assumed that the direction of the lenticules is vertical (except Chapter 7).

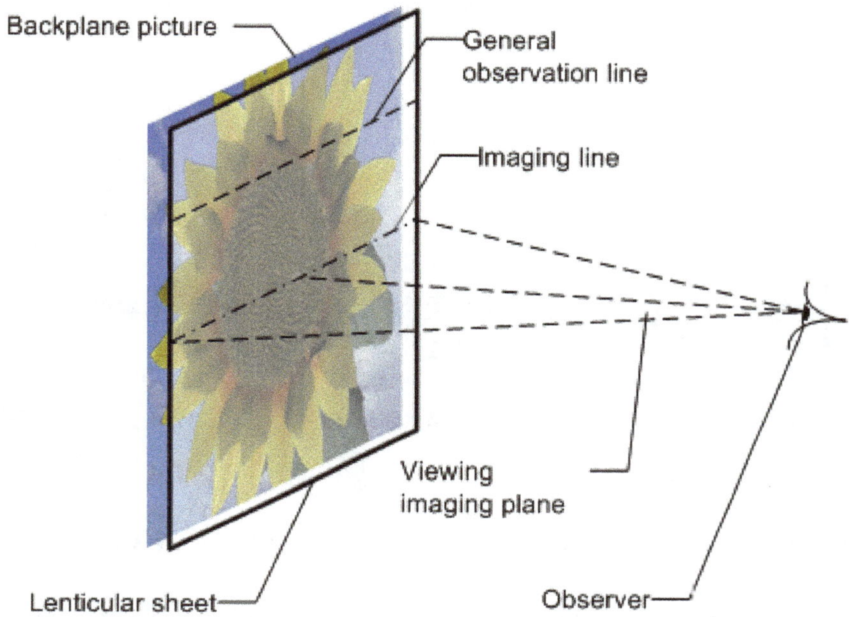

Figure 10: Viewing geometry of a lenticular picture.

The rectangle with the thick borders represents the lenticular sheet. It is displayed as a partially transparent window; the lenticules are not shown. The backplane picture is shown with a small gap behind the lenticular sheet. In reality, it is bonded to the backplane or deposited on it. Special backplane pictures can be used to enable the lenticular picture to display a different image for different observation points. In the present example, the backplane picture is a conventional one.

All viewing imaging planes of the lenticules in the sheet are identical, and therefore they can be referred to as the viewing imaging plane of the sheet. Like the viewing imaging plane of the lenticule, its position depends on the position of the observation point. The intersection of the viewing imaging plane and the sheet backplane is the imaging line. Any line parallel to it in the picture backplane will be called an "observation line."

2.8 The effect of a lenticular sheet

2.8.1 Imaging line display

When a lenticular sheet is placed on a picture, its visual appearance changes. Let us start the discussion of this effect by considering the relatively simple case of the imaging line.

The scheme of the viewing imaging plane shown in Figure 11 will be used to understand the visual effect of the lenticular sheet in the viewing line. The dotted lines are the axial rays, which intersect the backplane at a series of equidistant points, which are the axial points.

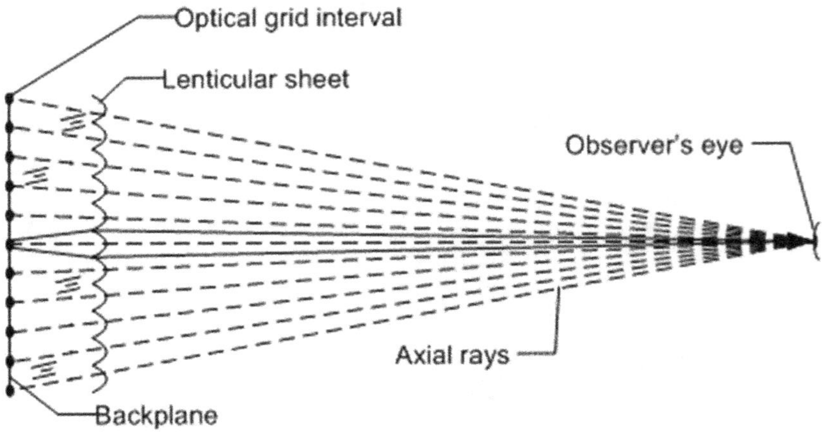

Figure 11: The optical effect of a lenticular sheet in the viewing imaging plane.

The optical gridlines intersect the imaging plane in a series of intervals, shown in Figure 11 as thick short segments on the backplane and annotated as "optical grid interval." Within the simplified model, the centers of these intervals are the axial points. Only the content included in the optical grid intervals contributes to the display on the imaging line.

Classical optical imaging analysis can be applied to derive the optical effect of the sheet in the imaging plane. The backplane of a lenticular sheet is designed to be also the focal plane, so the optical image of any object disposed on the backplane will appear at infinity. In practically all viewing scenarios, lenticular pictures are observed from distances which are much greater than the lenticule focal length, so that the observer may be considered as being at infinity. Therefore, the discussion concerning viewing from the image plane presented in paragraph 2.6 can be applied to the imaging plane in lenticular picture viewing.

The graphical content sampled by an optical grid interval is displayed over the whole aperture of its corresponding lenticule, so the observer sees the imaging line as a continuous display even though the optical grid intervals have wide gaps between them. This effect is illustrated in Figure 11 for the central lenticule only, in order not to obscure the figure. The graphical content displayed in each lenticule will be some average of the graphical content sampled from the backplane image by the corresponding optical grid interval. Thus, in the imaging plane, the size of the visual resolution element is the lenticule pitch, regardless of the printing resolution of the backplane picture.

The optical effect of a lenticule is illustrated with a straight line parallel to the lenticule axis. Figure 12 shows a photograph of a broken line covered by a small array of lenticules, and it was taken from a position in which the visible zone of one of the lenticules overlaps the line. The line itself is thinner than the lenticule aperture. The transverse magnifying effect of the lenticule is clearly noticed. This photograph also illustrates the fact that in the direction parallel to the lenticule axis, the geometry of the graphical content is preserved. Namely, the apparent lengths of the broken line segments are unaffected.

Figure 13 shows the same object photographed from a slightly different position, in which the visible zone lies outside the line. In this case, the lenticule makes the line disappear.

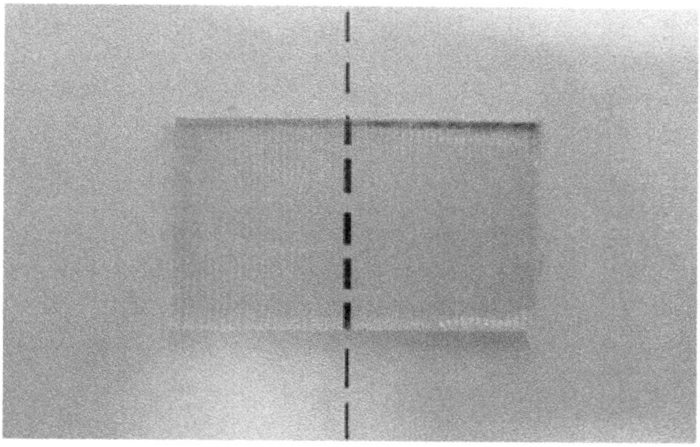

Figure 12: The effect of a lenticule on a broken line: the visible zone overlaps the line.

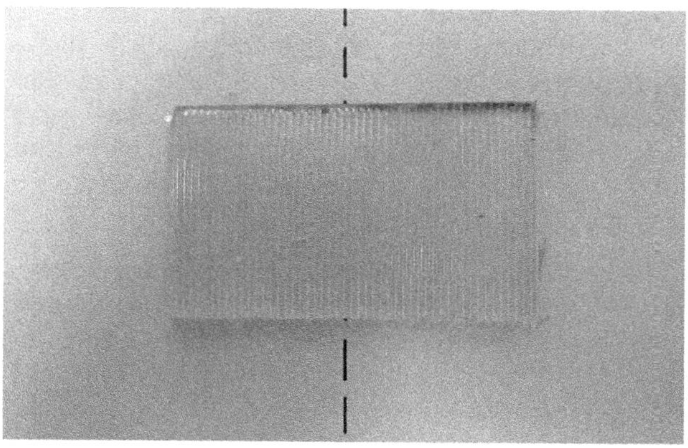

Figure 13: The effect of a lenticule on a broken line: the visible zone does not overlap the line.

Qualitatively speaking, the lenticular sheet behaves like an optical window in the axial planes, and as an array of imaging lenses in the imaging plane.

2.8.2 General observation line display

A rigorous analysis of general observation lines requires elaborate ray-tracing analysis. Such analysis will be presented in Chapter 11. Here, the visual effects in a general observation line will be described qualitatively.

The graphical content which contributes to the display of the imaging line comes from the optical grid intervals. The graphical content which contributes to the display of a general observation line comes from a series of spots. The width of these spots is equal to the optical gridlines width. Their height depends on the distance of the observation line from the imaging line, increasing with distance. This phenomenon is caused by lens aberrations and is evaluated quantitatively in Chapter 11. The finite height of the spots causes vertical blurring, which effectively limits the picture optical resolution in the vertical direction. This resolution limitation is, however, much less severe than in the horizontal direction, and is usually overlooked.

2.8.3 The apparent pixels

Printed pictures are made up from an array of small rectangles of uniform color, called "pixels." The pixel concept will be discussed in some detail in the next chapter. The optical properties of lenticules lead to a new concept: "apparent pixels." These are the pixels of the lenticular picture, which are different from the pixels of the printed backplane picture.

As explained in paragraph 2.8, the horizontal resolution element of the lenticular picture is lenticule pitch p, regardless of the resolution of the backplane picture, while the vertical resolution element is the same as in the backplane picture. Thus, the width of the apparent pixels is equal to the lenticules width, and their height is equal to the printed pixel height. Since usually the printed pixel size is much smaller than the lenticule width, several printed pixels are merged horizontally to form a single apparent pixel, as illustrated in Figure 14. This optical mechanism reduces the lenticular picture resolution in the horizontal direction, compared to the resolution of the backplane picture. Figure 14 shows only a small excerpt of the lenticular picture, which consists of two apparent pixels columns. In the full picture, the number of apparent pixels columns equals the number of lenticules.

Since the optical grid moves with the horizontal movement of the observer's eye, the content displayed in the apparent pixels will be different for different positions of the observer. However, for a given observation point, each apparent pixel will appear with a uniform color. The geometry of the apparent pixels grid remains invariant with respect to the observer's position.

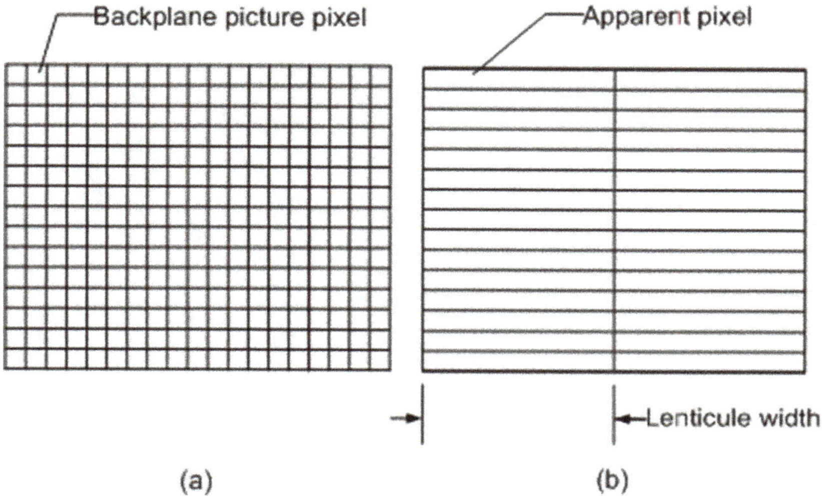

Figure 14: The printed pixels (a) and the apparent pixels (b).

Figure 15 illustrates how a lenticular sheet affects the appearance of the backplane picture. A sheet of only 50 lenticules was chosen so that the presence of the apparent pixels is very conspicuous. Typically, the number of lenticules in a lenticular picture is much higher, and therefore the visual effect caused by the loss of resolution in the horizontal direction is less pronounced.

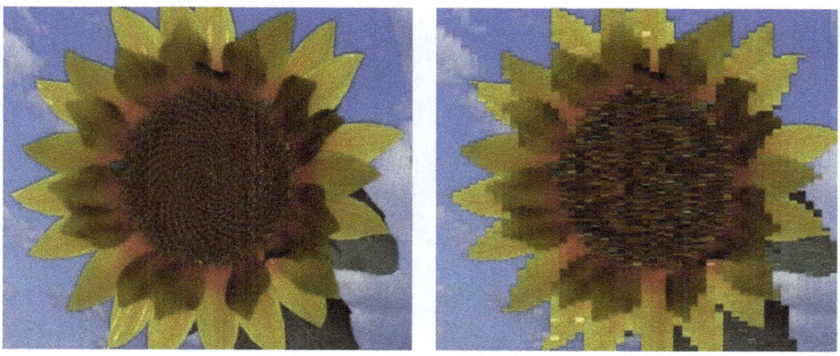

Backplane picture

Backplane picture with superimposed 50 lenticules sheet

Figure 15: The effect of a lenticular sheet on the backplane picture.

Chapter 3
Digital images, pictures, and transformations

"...the hardest rock will be worn away. Nothing endures but words"

- Robert Harris, in "Conspirata: A Novel of Ancient Rome" (attributed literary to Tiro, Cicero's secretary)

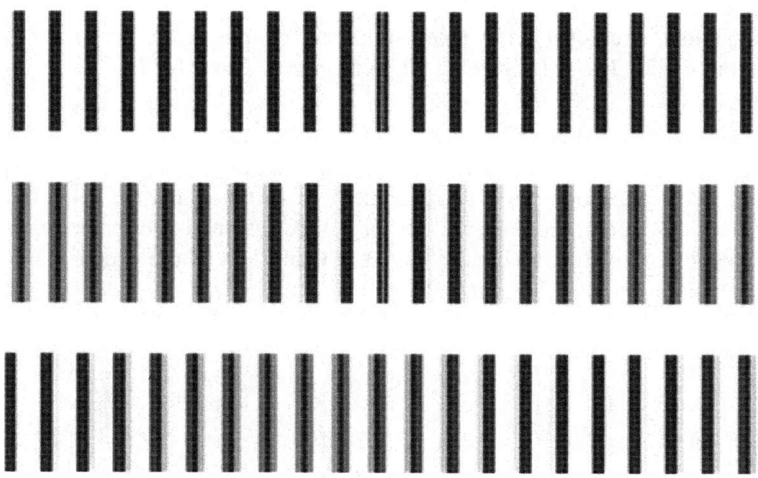

3.1 Images and pictures

With the advent of digital technology, it became possible to represent pictures as digital information. Such representations will be called "digital images," or in short, "images." An image can have a physical representation in the form of print. The print is most commonly paper with the image imprinted on it with ink. In this book, both representations will be used intermittently, and to avoid confusion, the physical representation will always be referred to as a "picture."

Images are defined on a rectangle. A coordinate system must be introduced to treat images mathematically. It is customary to choose the coordinate axes to be parallel to the rectangle edges, and the origin to coincide with the rectangle center, as shown in Figure 16. The coordinates do not have physical units so that the only significant geometrical property of the image rectangle is the ratio between width and height, which is called the "aspect ratio." In Figure 16, a unit of length was assigned to the image height, in which case the width obtained the value of the aspect ratio.

At least three digits represent the graphical content of a colored image point. For most purposes three digits are sufficient, but sometimes more than three are used to comply with various standards. In the present discussion, it will be assumed that the graphical content is represented by three digits representing red, green, and blue (R, G, and B) colors levels. The color levels will be denoted by a vector $\boldsymbol{\Gamma}$:

$$\boldsymbol{\Gamma} = [R\,G\,B].$$

Many color systems use 8-bit integers for the color levels values. In such systems, the color levels take values between 0 to 255. The representation and the manipulation of color are out the scope of the present book. It will be assumed that the color values are continuous and confined to the interval between 0 and 1, with the vector [0 0 0] representing black and [1 1 1] representing white.

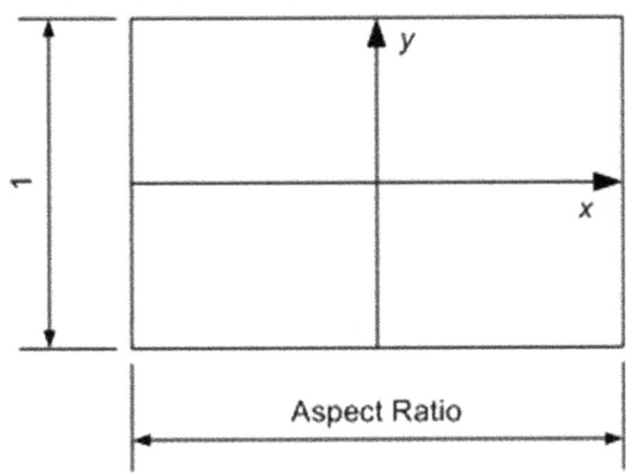

Figure 16: The image plane and the image rectangle.

3.2 Continuous and raster images

A continuous image (Pratt, 1991) is a color function $\boldsymbol{\Gamma}(x,y)$ defined on the image rectangle. Such function assigns the R, G, and B values to every point in the image rectangle. The color values can be different for any two given points, no matter how close they are.

For example, consider an image rectangle consisting of the points $[x\ y]$ such that both x and y are confined to the interval [-1 1], and the color function:

$\boldsymbol{\Gamma}(x,y) = [0.5\ 0.5\ 0.5]$ for $x^2 + y^2 < 1$, and

$\boldsymbol{\Gamma}(x,y) = [1\ 1\ 1]$ for $x^2 + y^2 \geq 1$.

The image corresponding to this color function is shown in Figure 17.

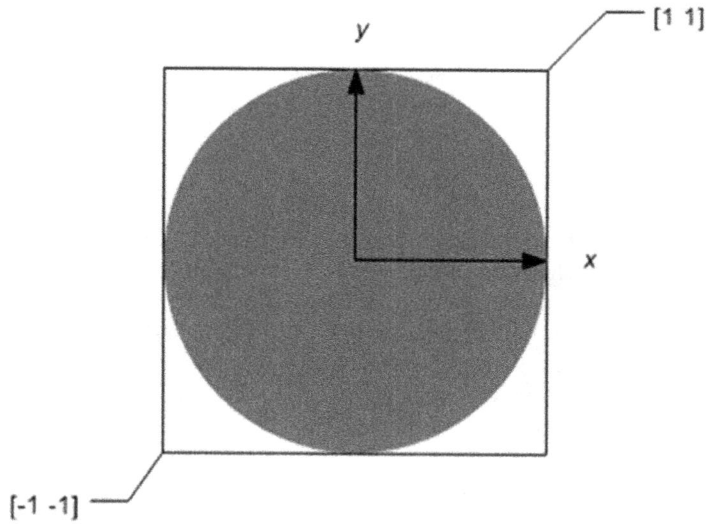

Figure 17: An example of an image defined by a continuous color function.

Readers familiar with computer graphics may classify this representation as "vector format" (Vector Graphics, 2017). This format uses various geometric shapes to define the color function, including circles such as in Figure 17. It is close to the continuous image concept, but not equivalent to it, because it has built-in geometric constraints, while the continuous image concept applies to images with arbitrary geometry.

Image capturing and display devices work with an image representation called "raster format." In this format, the image is represented by a two-dimensional array of rectangles called "pixels," with each pixel having a uniform color. Formally, this representation consists of three two-dimensional arrays (matrices), one array per color. By definition, any two image points which belong to the same pixel have the same color.

Most commonly square pixels are used, although this is not mandatory, and one can use pixels with any aspect ratio for a raster image. Allowing non-square pixels creates a certain redundancy. To avoid confusion, from now on all pixels will be assumed to be squares.

Modern digital cameras acquire the image with sensors which are two-dimensional arrays of photocells. The output of such sensors is a raster image. The pixel grid is defined by the number of pixels in each dimension. A common raster image format is "HD 1080 lines", with pixel dimensions of 1920 (width) x 1080 (height), which makes a total of 2,073,600 pixels. This format corresponds to an aspect ratio of 16:9, which is common in TV displays. Many other standard formats exist, but in principle, a raster image rectangle can have arbitrary pixel dimensions.

Pixels constrain the definition of graphical objects on a raster image. For instance, the smallest "dot" which can be defined on such a picture is a single pixel. Any horizontal or vertical line with a uniform color appears with a width larger than a single pixel. On the other hand, no such restrictions exist in continuous pictures, where one can define objects with any geometry, including pure (zero area) geometrical points and zero-width lines.

There are two principal images in lenticular imaging: the interlaced and the calibration images. These images are synthetic, in the sense that they are not a direct product of an image capturing device like a camera. Instead, they are created by software using special algorithms. It is most convenient to discuss these images in the continuous format. Such discussion avoids completely the technical complications which arise from the presence of pixels in the raster format and can focus on the essence. The use of the continuous format serves presentation purposes only; it does not make the technical complications disappear. These complications will emerge in real algorithms which must yield raster images as their output.

3.3 Conversion of a continuous image to a raster image

3.3.1 General

The process of converting a continuous image into a raster image is called raster rendering, or "rasterization". This process consists of two stages:

1. Pixelization,
2. Sampling.

Generally, rasterization involves a loss of information, and the rasterized image is different from the original vector image.

A continuous image must be rasterized before it is printed. There are many rasterization algorithms available (Wolberg, 1990) (Foley, Andries, Freiner, & Hughes, 1990) (Pratt, 1991). In general, a rasterization algorithm is judged by how well the raster image represents the original continuous image. There is no universally optimal rasterization algorithm because the outcome of rasterization depends on the image's graphical content. Thus, a given algorithm A may outperform an algorithm B for a given image X, but B can outperform A for another image Y. Nevertheless, the standard algorithms yield nearly optimal results for most images with conventional graphical content.

The graphical content of the interlaced and the calibration images is highly non-conventional, and it possesses certain unique graphical characteristics. Preservation of these characteristics is important for the quality of the lenticular picture, and this can be achieved in certain cases by pixel grid alignment. This operation will be described below, and its significance illustrated with several examples.

3.3.2 *Pixelization*

Pixelization is the definition of the pixel grid, which must contain a whole number of pixels in each direction. Most commonly, this grid is defined by assigning a certain whole number of pixels to one of the dimensions of the image. Let us assume that N_w pixels were assigned to the image width. The number of pixels in the height dimension will be

$$N_h = \left\{ \frac{N_w}{a} \right\}, \qquad \textit{Equation 13}$$

where a is the aspect ratio of the image and $\{x\}$ is a function which converts x to an integer. Naturally, one would like to choose a function which gives an integer with a value close to x. Three choices for this function are commonly used:

1. The smallest integer which is larger than x (sometimes called “ceiling” function),
2. The largest integer which is smaller than x (sometimes called “floor” function),
3. The closest integer to x (the “rounding” function).

Alternatively, it would be possible to choose the number of pixels N_h in the vertical direction and then determine N_w using the formula

$$N_w = \{aN_h\}\ .$$

Since, in general, $x \neq \{x\}$, the pixel grid rectangle will be slightly different from the original continuous image rectangle. Rasterizing programs usually do not disclose how this detail is handled; most probably the continuous image rectangle is somewhat resized to fit the raster grid size. Since the number of pixels in each dimension is ordinarily high (hundreds or thousands), this resizing has a negligible visual effect in conventional images.

Once the pixel grid is defined, the pixel size sets a virtual length unit, which can be used to measure various features in the continuous image. The dimension of a feature in this unit will be referred to as its pixel size. In general, the pixel size is not necessarily an integer. However, as explained above, in standard rasterization algorithms the pixel sizes of the image rectangle edges are made to be whole numbers.

3.3.3 Sampling

Sampling is the process of assigning colors to the pixels (Pratt, 1991). The goal is to represent the original continuous image as faithfully as possible in the raster image.

A simple and commonly used sampling method is to choose the color of the original continuous image point corresponding to the pixel center. A more elaborate method is to select the average color corresponding to the continuous image area which overlaps the pixel. Without going into elaborate details, it can be intuitively appreciated that this method will give better results in most cases.

Color averaging requires extensive computation for large images. Therefore, it is not implemented in most image processing programs, which use various interpolation algorithms instead. Such algorithms define the color of a given pixel by interpolating the colors corresponding to its center and the centers of the neighbor pixels. Popular image processing programs offer interpolation methods like nearest-neighbor, linear, bi-cubic, etc. Many more different sampling algorithms exist (Pratt, 1991) (Wolberg, 1990).

Simply sampled raster images often exhibit jagged edges, which are visually disturbing. Smoothing out this phenomenon is called in the image processing discipline “antialiasing” (Foley, Andries, Freiner, & Hughes, 1990), and it involves edge blurring. Virtually all practical image sampling algorithms perform antialiasing. In general, the blurring width is of the order of the pixel size. Therefore, to reduce the relative blurring of a given object, it is advisable to have its pixel size as large as possible.

3.4 Grid images

In the present context, a grid is a horizontal array of vertical rectangles (“columns”) with a uniform color. The rectangles height is the same as the image height, they are equally spaced and have the same width. A grid image is an image composed of grids with the same pitch. An example of a grid image is shown in Figure 18. This image is composed of two interleaved grids, one gray, and the other white. The columns of the grids in Figure 18 have equal widths.

If all columns in a given grid image have the same width, it is possible to define a special pixel grid, which is called "matched." In a matched pixel grid, the columns vertical boundaries coincide with the pixel gridlines. Such grid is shown in Figure 19. The matched grid is not unique; a grid of pixels half the size of a matched grid will be matched too, etc.

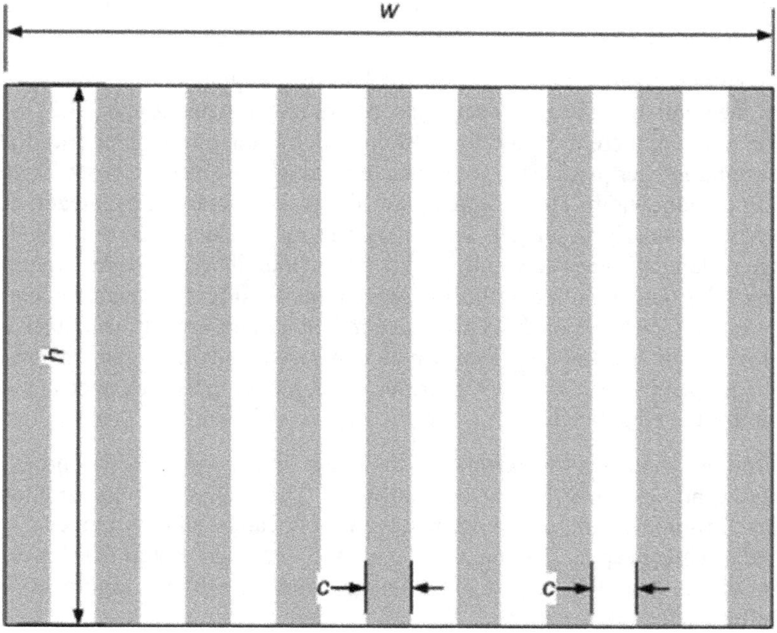

Figure 18: A grid image made of an array of alternating gray and white columns of equal width.

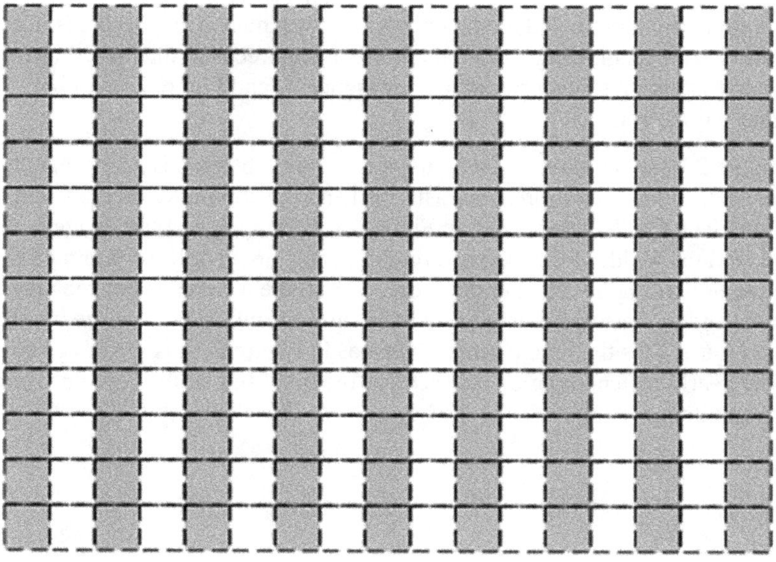

Figure 19: The rasterized image version with a matched pixel grid.

In general, a raster image is different from the original continuous image. Rasterization of a continuous grid image with a matched pixel grid yields a raster image which is identical to the original continuous image.

The common pixelization procedure involves the assignment of a whole pixel size to one of the continuous image dimensions, as explained in paragraph 3.3.2. In lenticular imaging, there is a need for a different definition of the pixel grid. Rather than assigning integer pixel sizes to the image rectangle, a given pixel size is assigned to the grid column width. The assigned pixel size is in general not an integer, but in many cases, it is nearly an integer. This creates a situation in which the pixel size of the image rectangle is not necessarily an integer in both dimensions. As explained in paragraph 3.3.2, in common pixelization procedures the original image rectangle is slightly resized to achieve a whole pixel size in both dimensions. In lenticular imaging, horizontal resizing is often forbidden, which may cause both image dimensions to have a non-integer pixel size. This is not a serious issue; it only calls for a special sampling algorithm for the edge pixels.

Let us now reconsider the example of the image shown in Figure 18 and assign a pixel size slightly less than 1 to the column width. This makes the pixel grid to extend beyond the original continuous image rectangle, so it can be slightly moved horizontally while keeping the image rectangle fully covered. Figure 20 shows a pixel grid position centered with respect to the image rectangle, while in Figure 21 the pixel grid is left-adjusted relative to it.

Whenever there is a gap between a column border and the nearest pixel gridline, the rasterization will cause its graphical content to mix with an adjacent column to a certain degree. In this case, the mixture will be between gray and white colors so that a white column will darken, and a gray column will brighten. This mechanism is perceived visually as blurring. It can, therefore, be appreciated that the rasterized images corresponding to the grids shown in Figure 20 and Figure 21 will be significantly different.

In Figure 20, the region marked by the five central columns is nearly matched to the pixel grid, so that in this area the rasterized and the continuous images will be almost identical. On the other hand, in Figure 21 the left edge of the image is nearly matched, and the width of the matched region is only about half the width of the matched region in Figure 20, and, therefore, overall the difference between the continuous and the rasterized versions will be more significant. Also, the unaffected region in Figure 20 is the image center, whereas in Figure 21 it is the image edge. The size of the nearly matched region increases as the pixel size of the column width is closer to an integer.

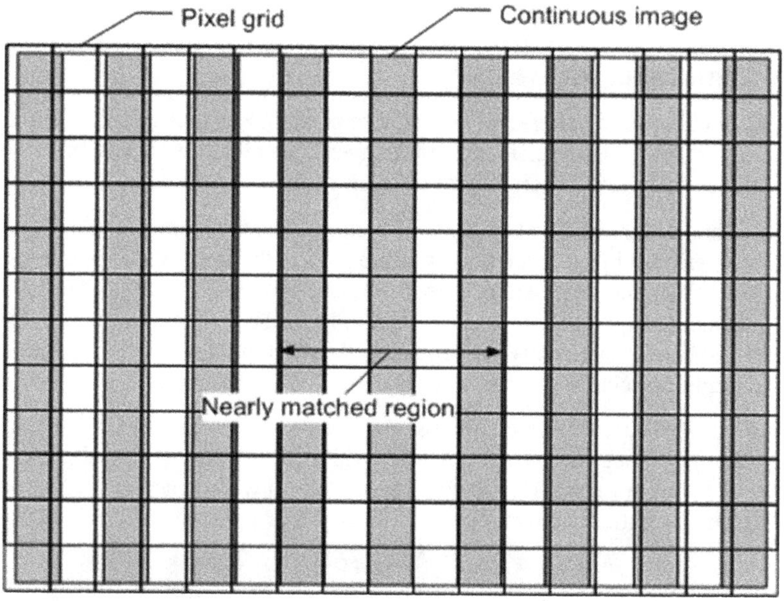

Figure 20: Centered nearly matched pixel grid.

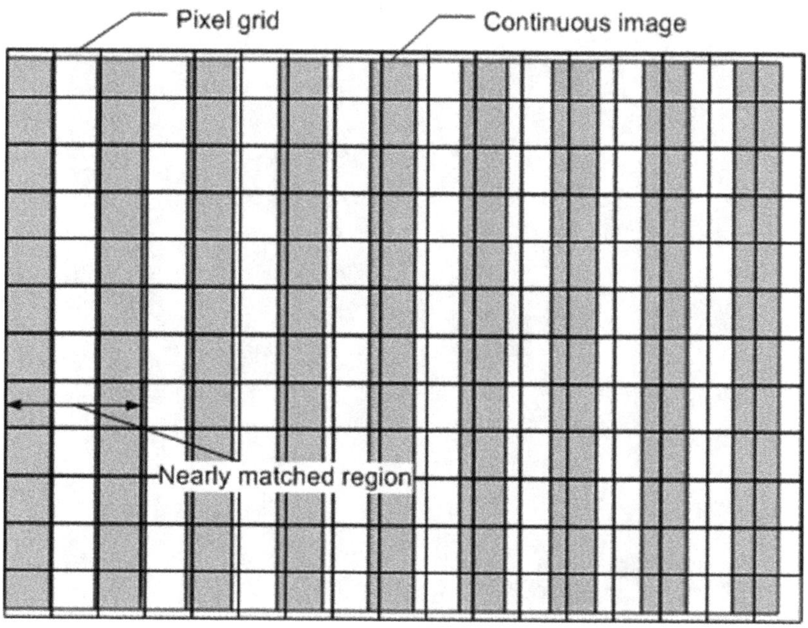

Figure 21: Left-adjusted nearly matched pixel grid.

This example illustrates the importance of aligning nearly-matched pixel grid to a grid image.

3.5 Cellular images

The standard calibration pattern, one of the two principal images of lenticular imaging, is a vertical stack of slightly different grid images. The other key image, the interlaced image, belongs to a more general class: cellular images.

Cellular images are made of an array of rectangular area elements called cells. In the case of interlaced images, the height of the cells is the same as the image height. All cells have the same dimensions. The difference between cells and pixels is that cells may contain graphical content, and their color is not uniform, although it is continuous. The difference between cellular and common images is the existence of color discontinuities across cells boundaries.

Cellular images are synthetic, in the sense that they are created by mathematical algorithms, not by image capturing devices. Grid images with equal-width columns, like the one shown in Figure 18, are cellular images, in which each grid column may be considered as a cell. A general example of a cellular image is shown in Figure 22.

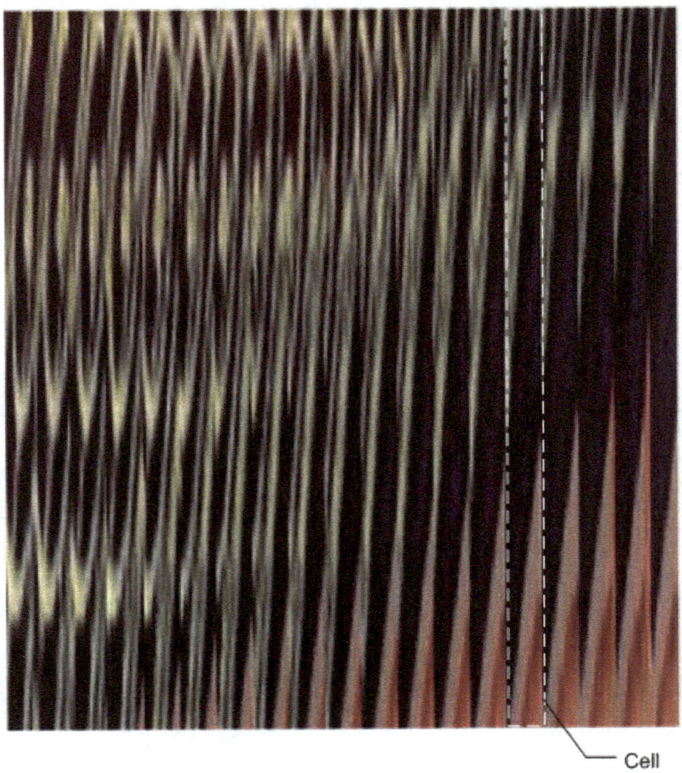

Figure 22: A cellular image

The concept of the matched pixel grid can be applied to cellular images too. A pixel grid is matched to a cellular image if all its pixel gridlines coincide with cells boundaries. In general, rasterization causes mixing of graphical content between different cells. When the pixel grid is matched, rasterization will not mix the graphical content between adjacent cells, and its only effect will be the pixelization and sampling of the graphical content within each cell independently. Such pixelization has little effect on the lenticular picture. On the other hand, mixing of graphical content between adjacent cells causes blurring. With a matched pixel grid, there is no mixing of graphical content between different cells, and therefore no rasterization blur.

3.6 Examples of the rasterization blur

Figure 23 is an illustration of the rasterization effects on a grid image made of an array of black columns, separated by a white gap twice their width. The top picture in Figure 23 shows a rasterized version of this image with a column width of precisely two pixels. The pixel grid of this image is matched to the original continuous image, and therefore both images are identical, and the rasterized image exhibits sharp edges.

The rasterized version of the continuous image with a column pixel size of 2.02 and a centered pixel grid is shown at the central band of Figure 23. The image center is little affected, and the rasterization blurring effect builds up as the image edges are approached.

The bottom band of Figure 23 shows rasterization with the same column pixel size, but with the pixel grid adjusted to the left edge. In this case, the left side of the image appears sharp (unaffected), and the center becomes blurred.

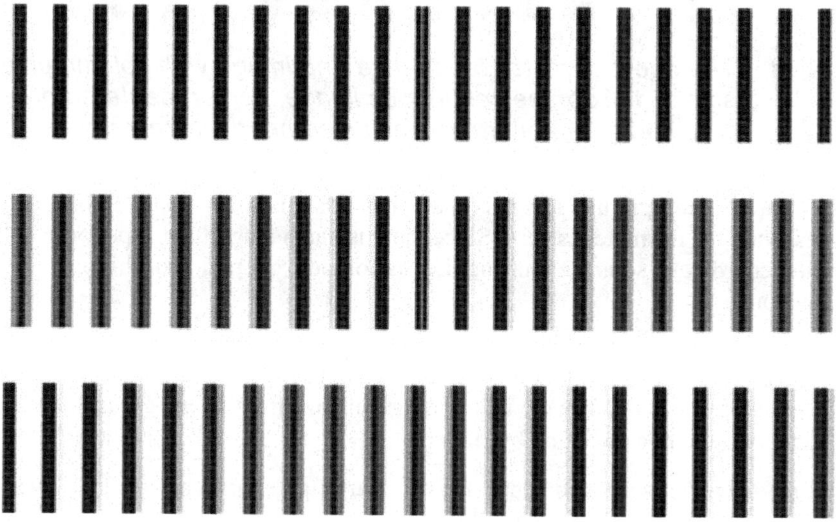

Figure 23: The effect of rasterization on a grid image with column pixel size of 2. Top: the continuous image, center: centered nearly matched grid, bottom: left-adjusted nearly matched grid.

Figure 24 shows a similar example with a single pixel assignment to the column width. In the top picture, the pixel size of a column is 1, and in the two pictures below it, it is 1.02. The blurring in Figure 24 is much more severe than in Figure 21, in agreement with our discussion in paragraph 3.3.3.

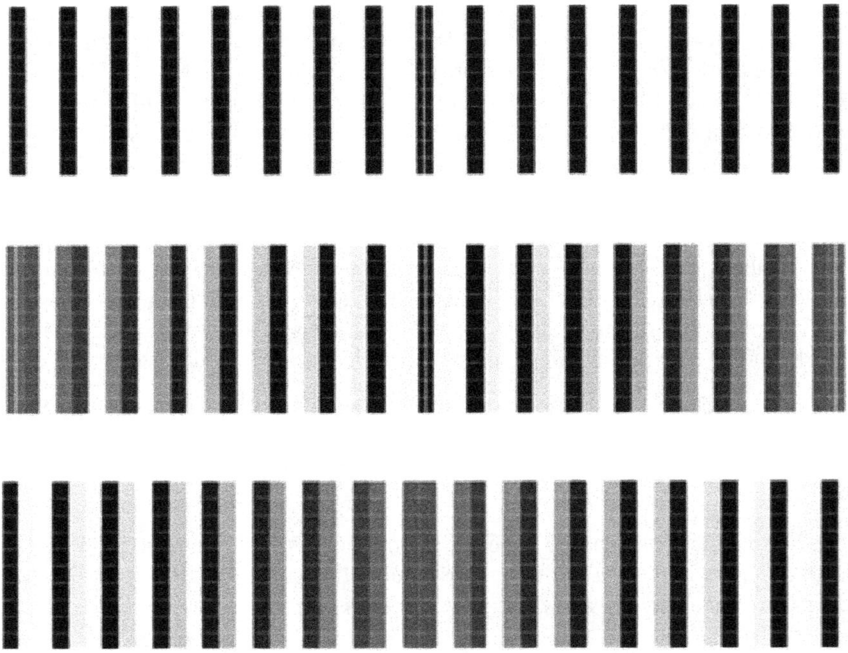

Figure 24: The effect of rasterization on a grid image with column pixel size of 1. Top: the continuous image, center: centered nearly matched grid, bottom: left-adjusted nearly matched grid.

The amount of blurring of a given column depends on the mismatch between its edge and the nearest pixel gridline. It is maximal when this mismatch is half a pixel, and it vanishes when the mismatch is zero. Since the mismatch magnitude is periodic in the horizontal coordinate, so is the blurring. Let us compute the period of this phenomenon.

The images in this example are periodic, with the cell width equal to three times the column width. In Figure 23, the pixel size of a column is 2.02, and the corresponding pixel size of a cell is 6.06. The deviation of the cell size from an integer is therefore 0.06. Let us denote this deviation by Δ.

The deviation between a cell edge and the nearest pixel line changes by Δ over a cell. In N cells, the mismatch will be changed by $N\Delta$. The number of cells U in which the mismatch changes by a whole pixel is the period of the blurring phenomenon. It is given by

$$U = \frac{1}{\Delta}$$

In this example $\Delta = 0.06$, and therefore $U = 16.7$. This period value can be verified by inspecting the bottom strips of Figure 23 and Figure 24.

It is of interest to evaluate how many columns will appear nearly sharp to the left and the right of the center column in a centered pixel grid, like in the central picture shown in Figure 23. This number can be estimated as a quarter of the blurring period. Thus, in the case of Figure 23, where $U = 16.7$, approximately four columns to the left and the right of the central column appear sharp. Thus, the number of nearly sharp columns U_C is

$$U_C = \frac{1}{2\Delta} \quad \text{Equation 14}$$

In the example shown in Figure 24, the value of Δ is the same as in Figure 23, namely 0.06. Correspondingly, U is the same too. This can be verified again by inspecting the bottom picture of the figure.

When U_C is larger than the picture width, the rasterization blur will be reduced over the whole picture. Also, by adjusting the pixel grid, it is possible to enhance the quality of a given region in the picture. For example, by using a centered grid, one may improve the quality in the picture center. On the other hand, if U_C is much smaller than the picture width, the position of the pixel grid has no significance.

In a matched grid, $\Delta = 0$.

3.7 Conversion of a raster image to a continuous image

The inverse process of converting a raster image to a continuous image is called "reconstruction" (Pratt, 1991). Reconstruction is not the mathematical inverse of rasterization. In general, applying both processes in sequence does not result in a return to the original image.

Since there are no constraints on the size of the continuous image, its rectangle can always be defined to be identical to the raster image rectangle. Thus, unlike in the case of the continuous to raster conversion, the aspect ratio in raster to continuous conversion is preserved.

A simple reconstruction algorithm consists of assigning a given point ***P*** in the image rectangle the *RGB* values of the pixel which contains it. Referring to Figure 25, according to this algorithm the *RGB* values of pixel ***B*** are assigned to the point ***P***.

Interpolation algorithms will usually achieve better results. Nearest-neighbor interpolation is illustrated in Figure 25. Point ***P*** is a general point in the image plane, and ***A***, ***B***, ***C***, and ***D*** are the nearest pixel centers. The *RGB* value at point ***P*** is computed by two-dimensional interpolation of the *RGB* values at points ***A***, ***B***, ***C***, and ***D***. Many more elaborate reconstruction schemes have been developed (Wolberg, 1990).

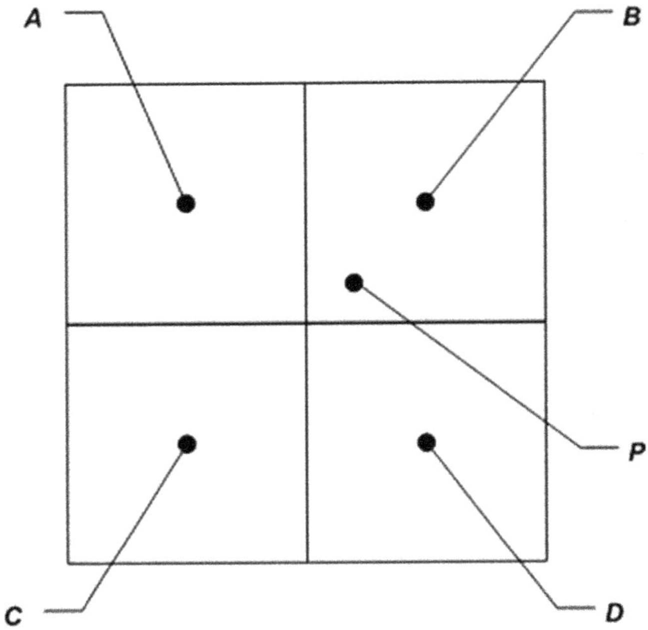

Figure 25: Nearest pixels restoration algorithm.

3.8 Geometrical transformations on continuous images

There are two types of image transformations: color and geometry. Only geometrical transformations will be discussed here.

Continuous images can be easily transformed into other continuous images. Let A be the rectangle of the original image, and A' the rectangle of the transformed image. Let T be a coordinate transformation which transforms a point $\boldsymbol{x}$ in A to a point $\boldsymbol{x}'$ in A':

$$\boldsymbol{x}' = T(\boldsymbol{x}) .$$

If the content of point $\boldsymbol{x}$ is transferred to point $\boldsymbol{x}'$ we get a geometrically transformed image. The new color function, which is denoted by $\boldsymbol{\Gamma}'$, is given by

$$\boldsymbol{\Gamma}'(\boldsymbol{x}') = \boldsymbol{\Gamma}\left(T^{-1}(\boldsymbol{x}')\right),$$

where T^{-1} is the inverse of T:

$$\boldsymbol{x} = T^{-1}(\boldsymbol{x}') .$$

Let us consider a simple example: a shift transformation. Such transformation is characterized by a vector $\boldsymbol{p}$, and the transformed point is related to the original point by

$$\boldsymbol{x}' = \boldsymbol{x} + \boldsymbol{p} .$$

The new color function becomes

$$\boldsymbol{\Gamma}'(x') = \boldsymbol{\Gamma}(x' - p) .$$

Another simple example is scaling transformation, in which the new and the original points are related by

$$x' = \frac{x}{\delta} ,$$

where δ is the scaling factor. In this case, the new color function is

$$\boldsymbol{\Gamma}'(x') = \boldsymbol{\Gamma}(\delta x') .$$

In both scaling and shifting, the image aspect ratio is conserved, and the transformed image looks exactly like the original one. A scaling transformation in a single dimension can be defined by

$$x' = \frac{x}{\delta} \quad y' = y . \qquad \textit{Equation 15}$$

Such transformation decreases the aspect ratio by a factor of δ, and the graphical content is distorted. For $\delta > 1$ we call this transformation "contraction"; otherwise, it is called "stretching". An example of a contraction transformation is shown in Figure 26.

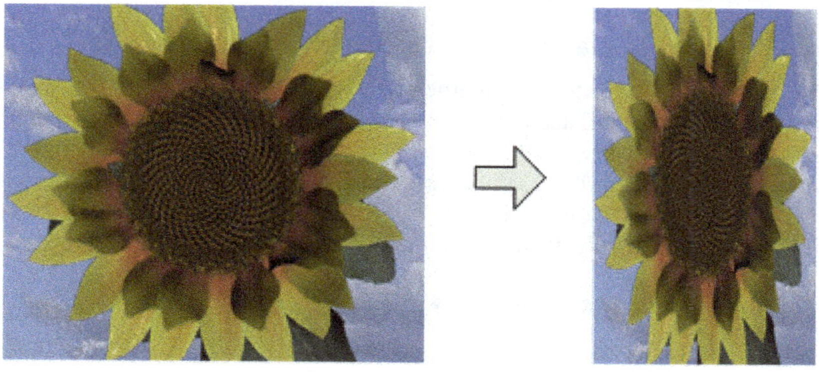

Figure 26: A contraction transformation with a factor of 2.

Both the shift and the scaling transformations map a rectangle to a rectangle. In general, the set of points $T(\boldsymbol{x})$ is not a rectangle; a typical example is the rotation transformation. In such cases, A' is chosen to be the rectangle which contains the set $T(\boldsymbol{x})$. Not all points in this rectangle will be assigned a graphical content by the transformation. Different image processing programs assign different default colors to blank regions of A'. Here, such regions will be shown in uniform black. When the image rectangle contains blank regions, it is customary to refer to it as a "canvas."

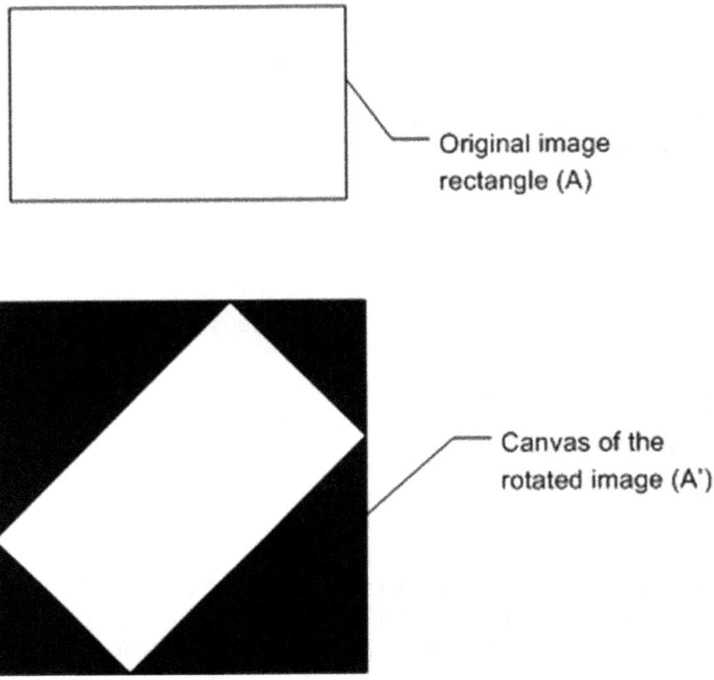

Figure 27: Original image and the canvas of the rotated image.

One may envision the transformation process by regarding the image as if it was printed on a thin elastic foil and deposited on the image rectangle, which is imagined as a stiff substrate. To carry out the transformation one separates the foil from the substrate, deforms it according to the specified transformation, and then bonds it back to the substrate. In some cases, like in the case shown in Figure 27, the substrate needs to be extended to contain the whole transformed image.

3.9 Projective transformations

Projective transformations (Hartley & Zisserman, 2003) (Wolberg, 1990) are geometrical transformations which preserve straight lines. This means that straight lines in the original image are transformed to straight lines in the transformed image. Angles are not necessarily preserved by a projective transformation.

Projective transformations play an important role in preparing the content for 3D lenticular pictures. Any two images acquired by a camera from the same point are related to each other by a projective transformation (disregarding lens aberrations). The use of projective transformations allows perfect registration of a sequence of photographs which is required to produce high-quality 3D lenticular pictures.

The mathematical equations defining projective transformations will not be presented here, as software for such transformation is readily available in virtually all main programming languages. Interested readers are referred to the literature (Hartley & Zisserman, 2003) for details.

A projective transformation can transform a given quadrilateral to any other quadrilateral. Moreover, two quadrilaterals uniquely define the projective transformation. Therefore, a projective transformation is uniquely determined by two sets of four coordinates in a plane: the four vertex coordinates of the original quadrilateral, and the four vertex coordinates of the transformed quadrilateral. This is illustrated in Figure 28.

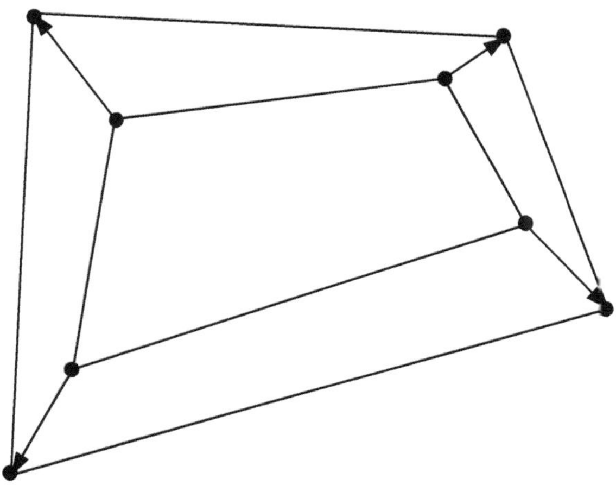

Figure 28: Example of two quadrilaterals which define a projective transformation.

3.10 Raster image transformations

A raster image transformation consists of three steps:

1. Conversion to a continuous image (reconstruction),
2. Applying the transformation,
3. Conversion of the transformed image to a raster image (pixelization + sampling).

The pixel grid of the final image may be different from the pixel grid of the original image. Steps 1 and 3 are accomplished as discussed in paragraph 3.7 and paragraph 3.3, and step 2 as discussed in 3.8. The following objects are needed to carry out a raster image transformation:

1. The input raster image (a file),
2. The transformation (a mathematical formula),
3. The pixel grid of the final image.

Chapter 4
The interlaced image

"It's incredibly frustrating to work so hard for no result. But very common in science."

- Graeme Simsion, in "The Rosie Project"

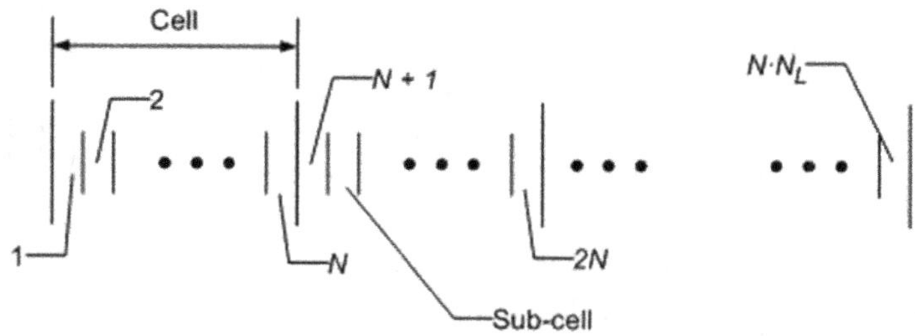

4.1 Geometry

Special backplane pictures create the unique effects of lenticular pictures, and they are called "interlaced pictures." The interlaced picture area is divided into an array of cells. There is a unique cell per lenticule, so the number of cells in the interlaced picture is the same as the number of lenticules in the sheet. The cell height equals the sheet height, and its width has a certain value p_p, which will be referred as the "printed grid pitch." When deposited on the sheet backplane, the interlaced picture is adjusted horizontally, so that it is centered with respect to it.

Let us assume that the sheet contains a non-even number of lenticules, so there is a central lenticule. The cell corresponding to the central lenticule will be called the "central cell." The lenticular sheet and the interlaced picture are shown in Figure 29 where the central cell is emphasized by a gray color. The lenticules are not shown in Figure 29, and the number of cells is vastly reduced with respect to the number of cells in a real picture for the sake of clarity. The graphical content of the interlaced picture is also not shown, and only its cells borders are outlined.

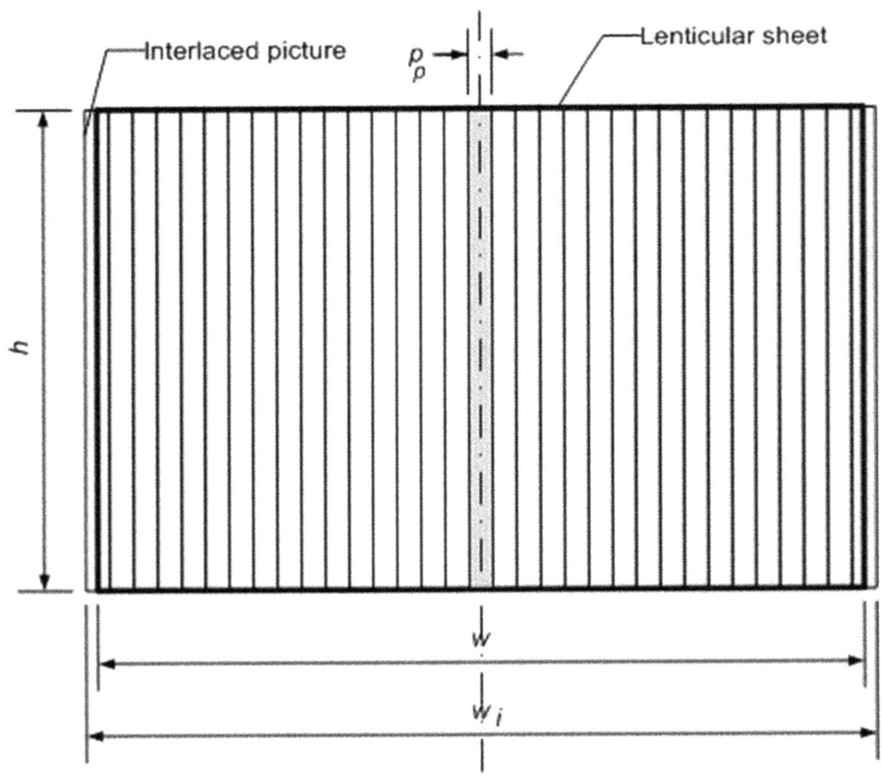

Figure 29: The lenticular sheet and the interlaced picture.

The interlaced picture width w_i is equal to $N_L p_p$. It will be seen below that the interlaced image is printed so that the grid pitch p_p is made equal to the optical grid pitch p_o for a given viewing distance. Since p_o is always slightly greater than the width p of the lenticules, the interlaced picture is slightly wider than the lenticular sheet.

4.2 The interlacing algorithm

The interlaced image is used to create a lenticular picture which will display different images for different observation points. These images are referred to as a "lenticular sequence." Since they are intended to be displayed by the lenticular picture, their aspect ratio should match the aspect ratio of the lenticular sheet.

The interlaced image is created from the lenticular sequence by using a special image processing algorithm. It is simplest to present this algorithm in the continuous representation. Once the interlaced image is created, it can be rasterized and printed to make the interlaced picture.

The input for the interlacing algorithm is the lenticular sequence and the number of lenticules N_L. First, a blank continuous image with an aspect ratio a_i is created:

$$a_i = \frac{N_L N p}{h} \quad . \qquad \textit{Equation 16}$$

Then, it is divided into N_L vertical cells, and each cell is further divided into N vertical sub-cells, where N is the number of the images in the sequence. The aspect ratios a_c and a_s of the cells and the sub-cells respectively are given by:

$$a_c = \frac{Np}{h} \; .$$

and

$$a_s = \frac{p}{h} \; .$$

A serial number is assigned to each sub-cell, as shown in Figure 30.

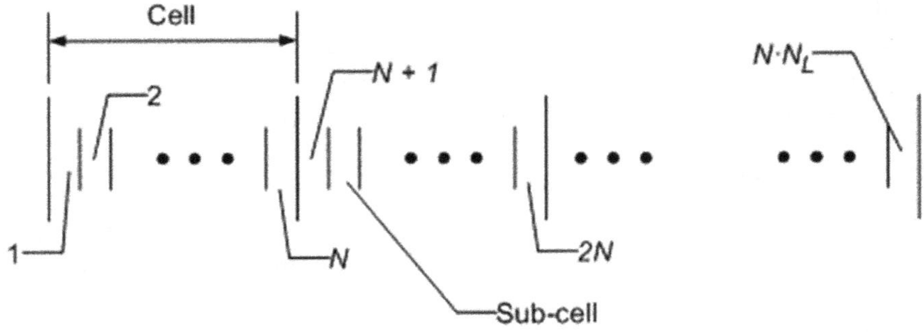

Figure 30: The structure of the interlaced image.

All images of the lenticular sequence are converted to a continuous format, and each one divided into N_L equal vertical cells. This conversion is necessary because raster images can be divided only along pixel gridlines, and here this limitation must be removed. Since the aspect ratio of the sequence images is $N_L p/h$, the aspect ratio of their cells is p/h, same as a_s.

The interlaced image is created by executing the following loop:

For $i = 1$ to N_L and $k = 1$ to N, map the content of the i-th cell of the k-th sequence image in sub-cell $(i - 1)N + k$ of the interlaced image.

Since the aspect ratios of the sequence image cells and the interlaced image sub-cells are identical, this mapping is an identity map (point to point), and as such does not require any further elaboration. An example of an interlaced image of a 3D picture is shown in Figure 31.

The interlaced image is cellular (paragraph 3.5), but with an added feature: cellular hierarchy. This hierarchy is manifested by each cell being composed of sub-cells. A pixel grid is matched to an interlaced image if it is matched to its sub-cells. However, it is possible to have a pixel grid matched to the cells, and not to the sub-cells. Such grid is a more general type of a matched grid, and will be called "periodically matched grid." A matched grid is also periodically matched; the opposite is not necessarily true.

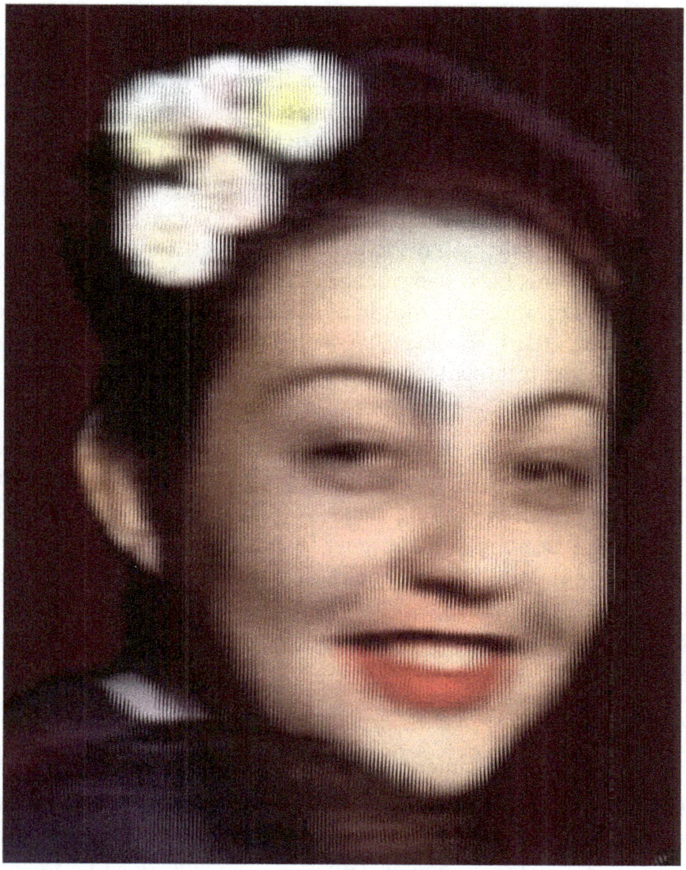

Figure 31: Example of an interlaced image.

4.3 Printing the interlaced image

4.3.1 Choosing the horizontal printing density

To print an image, one needs to specify the printing density ρ. The printing density denotes the number of pixels that the printer will print per unit of length. The printed pixel size s will be

$$s = \frac{1}{\rho} .$$

Common printing programs, as well as printer drivers, accept only one value of ρ, which is used for both picture dimensions. As explained below, in printing the interlaced image, it is beneficial to be able to specify different printing densities for the two image dimensions.

Traditionally, printing densities are specified with an inch as the unit of length. A printing density unit is called “dpi," which stands for "dots per inch." This name is misleading because in printing technology a dot has an entirely different meaning from a pixel. In fact, a printed pixel consists of many dots. The specification of the printing densities determines the physical sizes of the printed picture.

It is well known that digital printers have certain preferred printing densities, in which the print quality is better than in other densities. In conventional pictures, these differences in quality are usually too small to matter. This is not the case in lenticular printing. The author of this book had a chance to investigate digital printers only. The existence of preferred printing densities in digital printers was verified by him, as well as by others (Gardner, 2016). However, the author did not find any information on similar behavior in other printing technologies. More details on this subject are given below in Chapter 10.

Different digital printer models have different preferred densities. Moreover, a given model may have more than one preferred printing density. In particular, it appears that if ρ is a preferred printing density for a particular printer model, then so is a whole multiple of ρ. Leading digital printer brands feature different sets of preferred printing densities: multiples of 720dpi and multiples of 600dpi. It will be demonstrated below (Chapter 10) that the densities 720dpi or 600dpi are also approximately the limits of the printer's resolution. Therefore, nothing can be gained regarding printing quality by specifying higher printing densities.

4.3.2 *Matching the printed and the optical grids*

The cell width p_p in the printed interlaced picture must be matched to the optical grid pitch p_o. Since the optical grid pitch depends on the viewing distance, this matching may be accomplished only for a specific viewing distance V_0. In the simplified model, the optical grid pitch equals the axial grid pitch p_a, which is given by Equation 12. Therefore, the cell width of the interlaced picture is given by

$$p_p = p\left(1+\frac{t}{nV_0}\right) \qquad \textit{Equation 17}$$

This distance V_0 will be called “calibration distance." It can be derived from p_p by inverting Equation 17:

$$V_0 = \frac{pt}{n(p_p - p)} \ . \qquad \textit{Equation 18}$$

Let us denote the horizontal printing density by ρ_h. The pixel size c of the sub-cell width is:

$$c = \frac{\rho_h p_p}{N} \ . \qquad \textit{Equation 19}$$

Once the pixel size of the sub-cell width is determined, the interlaced image can be rasterized as explained in paragraph 3.5 and then printed. In general, c will not be a whole number.

4.4 The mechanism of the lenticular picture

The lenticular picture is made by depositing the interlaced picture on the lenticular sheet backplane, either by a direct print or by lamination. In both cases, the printed and the lenticules grids must be parallel, and the interlaced picture should be centered with respect to the lenticular sheet, as shown in Figure 29.

When the lenticular picture is viewed from the calibration distance, all optical gridlines will be contained in sub-cells corresponding to the same image from the sequence. When the observer moves horizontally, the optical grid moves correspondingly, and its gridlines will be contained in sub-cells corresponding to another image from the sequence.

The lenticular picture displays the image which is sampled by the optical gridlines. Therefore, for different observation points, different images will be displayed by the lenticular picture.

4.5 The pixel count of the interlaced image

The pixel size assignment of Equation 19 determines the pixel count w_p of the interlaced image width:

$$w_p = \{N_L N c\} \,, \qquad \textit{Equation 20}$$

where for the present discussion we define $\{x\}$ as the smallest integer larger or equal to x (ceiling function).

The interlaced image is often printed with the same vertical printing density as the horizontal. This sets the pixel count h_p of the image height to

$$h_p = \left\{ \frac{w_p}{a} \right\} . \qquad \textit{Equation 21}$$

The total pixel count N_p of such image is

$$N_p = w_p h_p \approx \left\{ \frac{(N_L N c)^2}{a} \right\} . \qquad \textit{Equation 22}$$

For large 3D pictures, this often turns out to be a huge number. Fortunately, the pixel count of the rasterized interlaced image can be significantly reduced without affecting the lenticular picture quality.

It was already explained that the visual resolution element in the horizontal direction is the lenticule width (2.8.3). In the vertical direction, the visual resolution element is the printed vertical pixel size, like in a conventional print. Experience shows that the overall visual resolution is determined by the largest resolution element. According to this, let us replace Equation 21 by

$$h_p = \left\{ \frac{rN_L}{a} \right\} . \qquad \textit{Equation 23}$$

where r is a number greater than 1. In most cases, choosing $r = 2$ is sufficient. The setting of h_p to the value given by Equation 23 will be called "collapsing." Combining Equation 20 and Equation 23, one gets the following result for the total pixel count of a collapsed interlaced image:

$$N^c{}_p = \left\{ \frac{rN_L{}^2 Nc}{a} \right\} . \qquad \textit{Equation 24}$$

Using Equation 22 and Equation 24 one gets

$$\frac{N_p}{N^c{}_p} \approx \left\{ \frac{Nc}{r} \right\} .$$

In 3D pictures, the ratio c/r is of the order of 1. Thus, the number of pixels in the collapsed format is reduced approximately by N, which is very significant.

Let us consider an example of a 1mx1m 3D lenticular picture made with a 20lpi lens. Such picture contains approximately 787 lenticules and has a 1:1 aspect ratio. A typical sequence length for such a picture is 30. The calculation of the number of pixels in the collapsed interlaced image for this case is presented in Table 5, and the result is about 74 megapixels. This is approximately the pixel size needed to print a 73cmx73cm picture at 300dpi, which is considered as high resolution. Thus, the number of pixels required for a lenticular picture of a given size is similar to the number of pixels required for a high-resolution print of a conventional image of comparable size, if the collapsed interlaced picture is used.

Table 5: Typical pixel count of an interlaced image of a large 3D picture.

Quantity	Symbol	Value
Picture width	w	1m
Picture height	h	1m
Lenticule density	-	20lpi
Sequence length	N	30
Sub-cell pixel width	c	2
Number of lenticules	N_L	~787
Number of pixels per lenticule width in the vertical direction	r	2
Aspect ratio of the lenticular picture	a	1
Total pixel count (Equation 24)	$N^c{}_p$	$\sim 74 \cdot 10^6$

The vertical printing density ρ_v of a collapsed picture is determined from the requirement that the printed picture height will be h:

$$h = \frac{h_p}{\rho_v} .$$

The printing quality requirements in the vertical direction are not critical, so ρ_v does not need to be a preferred density, so r can be chosen at will.

4.6 Printing with reduced rasterization blur

When the pixel size c of the sub-cell width is an integer, the pixel grid is matched. As explained in paragraphs 3.4 and 3.5, there will be no rasterization blur in such case. that assigning a non-integer value to c will cause rasterization blur. However, when the discrepancy Δc between c and the nearest integer is small enough, the blur may be reduced over a region of width z, as shown in Figure 32.

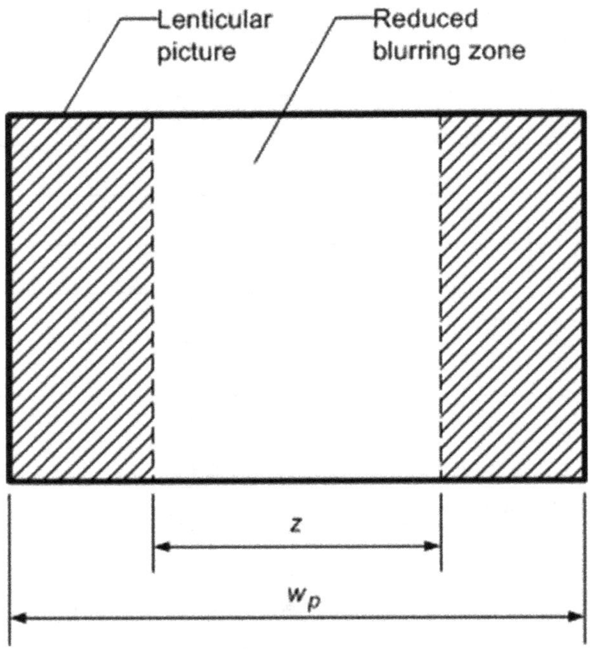

Figure 32: Reduced rasterization blurring for c nearly an integer.

The number of cells with reduced blurring is given by Equation 14:

$$U_c = \frac{1}{2N|\Delta c|} \ .$$

since the deviation Δ of the cell pixel size from an integer is $N|\Delta c|$. Correspondingly, the reduced blurring region width z is given by

$$z = U_c p = \frac{p}{2N|\Delta c|} \ .$$

The picture width is $N_L p$. Therefore, if

$$|\Delta c| < \frac{1}{2NN_L} \ , \qquad \textit{Equation 25}$$

the rasterization blur of the whole picture will be reduced.

Using the parameter values from the example of Table 5, the following condition is obtained:

$$|\Delta c| < 1.06 \cdot 10^{-5} \ .$$

This demonstrates that in order to benefit from reduced rasterization blur, the discrepancy of c from an integer value must be extremely small. This discrepancy is inversely proportional to the picture width (which is proportional to N_L).

Expressing the optical grid pitch p_o by the corresponding density ρ_o, Equation 19 implies that

$$c = \frac{\rho_h}{N\rho_o} .$$

Using this relation, values of N and ρ_o can be determined which give integer values of c, for a given printing density. Table 6 and Table 7 show such values of N for several optical grid densities, which are compatible with commercial lenticular sheet densities.

Table 6: Values of ρ_o and N for 600dpi printers which feature integer pixel size for the column width.

ρ_o (1/inch)	N
100	6, 3, 2
60	10, 5, 2
40	15, 5, 3
20	30, 6, 5, 3
10	60, 10, 6, 5, 4, 2

Table 7: Values of ρ_o and N for 720dpi printers which feature integer pixel size for the column width.

ρ_o (1/inch)	N
90	8, 4, 2
60	12, 6, 4, 3, 2
40	18, 9, 2
20	36, 18, 9, 4, 2
10	72, 36, 18, 9, 4, 2

The lenticule densities values must be slightly modified to account for the difference between the lenticule pitch p and the optical grid pitch p_o. According to Equation 12, this modification depends on the sheet parameters t and n, and on the viewing distance V.

Using the fact that the pitch is the reciprocal of the density, Equation 12 can be rewritten in the form

$$\rho_L = \rho_o\left(1+\frac{t}{nV}\right).$$ *Equation 26*

This equation can be used to compute the lenticule density ρ_L which is necessary to achieve a given value of ρ_o according to Table 6 and Table 7. A few such values are presented in Table 8, which was calculated for $n = 1.6$.

Table 8: Required values of lenticules densities ρ_L for several values of optical grid densities ρ_o.

ρ_o **(1/inch)**	t **(mm)**	V **(mm)**	ρ_L **(1/inch)**
100	0.5	500	100.0625
60	1	500	60.075
40	2	1000	40.05
20	4	2000	20.025
10	6	3000	10.0125

Some manufacturers design lenticular sheets to match the densities shown in the rightmost column.

4.7 Periodically matched pixel grids

A periodically matched grid can be defined for any lenticular sheet whose optical grid density is a whole divisor of a preferred printing density. For example, for an optical grid with density of 40lpi, one can define a matched grid by choosing $c = 18$ for printing density of 720dpi. If a sequence of 9 images is used, this grid may be matched to both sub-cells and cells. However, if a sequence of 10 images is used, the pixel grid may be matched to the cells only.

Periodically matched grid does not eliminate rasterization blur, but its blur affects displayed images rather than spatial regions in the picture. With a periodically matched grid, some images of the sequence may appear in the picture sharper than the others, and this fact may be used to optimize the picture visual quality.

Chapter 5
Calibration

"If you only read the books that everyone else is reading, you can only think what everyone else is thinking."

- Haruki Murakami, in "Norwegian Wood"

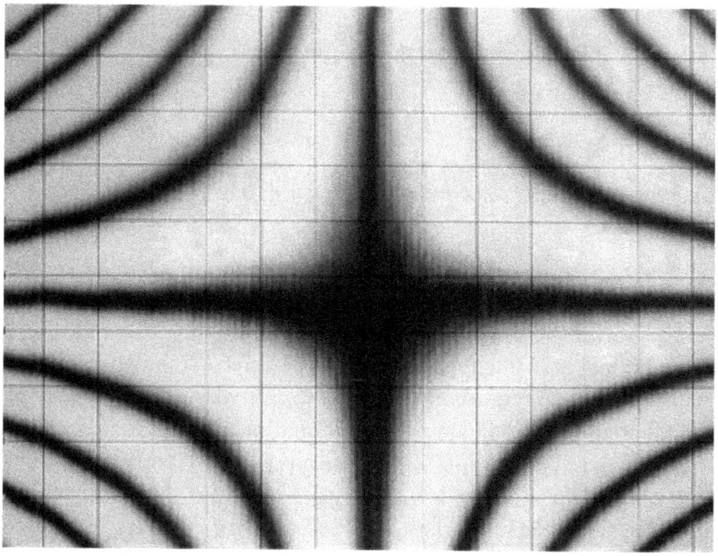

5.1 The accuracy requirement

In the present context, calibration is the process of measuring the optical grid pitch. Before making the measurement, it is necessary to specify the required accuracy.

A lenticular picture is designed for a specific viewing distance. With a perfect calibration, the specified and the calibrated viewing distances coincide. An error in the calibration measurement causes a small discrepancy between the two. Specifying the tolerance in the calibration distance determines the required accuracy in the calibration.

Within the simplified model, the optical and the axial grids pitch values coincide. Therefore, one can use formula Equation 12 to calculate the optical grid pitch. The change Δp_o in the optical grid pitch which is caused by a change ΔV in the viewing distance can be calculated using the well-known differential approximation

$$\Delta p_o = \frac{dp_o}{dV}\Delta V \ .$$

Using Equation 12, this can be written in the form

$$\Delta p_o = -\frac{tp}{nV^2}\Delta V \ . \qquad \textit{Equation 27}$$

Denoting the required relative accuracy $|\Delta p_o/p|$ by ζ, one can write

$$\zeta = \frac{t}{nV^2}\Delta V \ . \qquad \textit{Equation 28}$$

Let us consider an example with the values shown in Table 9. The required relative accuracy in the optical grid pitch measurement is $1.25\cdot 10^{-4}$ (shown in the last row of the table). This is a typical requirement, and it reveals why a direct measurement of p_o is necessary. An attempt to evaluate p_o using Equation 12 with such accuracy will require the value p to have the same relative accuracy ($1.25\cdot 10^{-4}$). This is normally unavailable. Even if the value of p is given by the manufacturer with such accuracy, it cannot be used because relative variations of the order of 10^{-4} may exist between different rulers. Therefore, a direct measurement of the optical pitch is required.

Let us note that the relative accuracy requirement does not depend on the lenticular sheet pitch p or the picture size.

Table 9: Example for calculation of the relative tolerance in the calibration measurement.

Parameter	**Symbol**	**Value**
Thickness	t	2mm
Index of refraction	n	1.6
Viewing distance	V	1000mm
Required tolerance in viewing distance	ΔV	100mm
Required relative tolerance (calculated)	ζ	$1.25 \cdot 10^{-4}$

5.2 The calibration pattern

5.2.1 General

A distinctive pattern is used to determine the optical grid pitch p_o. This pattern is a regular grid of black vertical lines on a white background and called a "calibration band." A compound calibration pattern can be made by stacking vertically an array of calibration bands, each having a different pitch. With such a pattern, a range of pitch values can be tested simultaneously with a single calibration pattern.

There are some variations and elaborations of this basic pattern, most notably the use of several multiplexed grids with different colors instead of a single black and white grid. Here only the basic method will be described, using a single black grid on a white background.

5.2.2 The calibration band image

The calibration band image consists of a horizontal array of identical cells. A scheme of a basic cell used to create a calibration band is shown in Figure 33. It is made up of two regions: black and white.

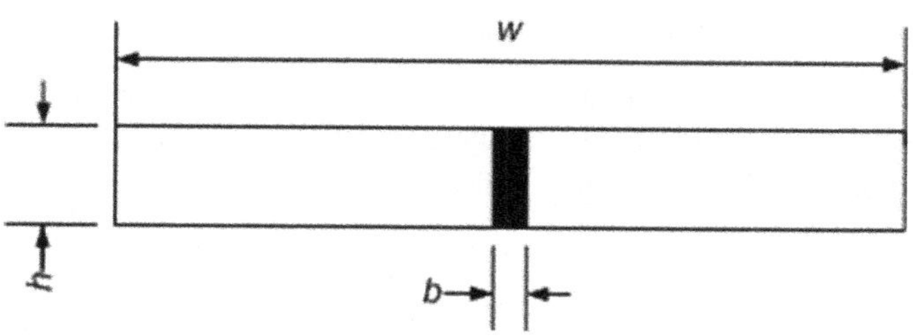

Figure 33: Calibration band cell.

Three parameters characterize the basic cell: its width, height, and the width of the black region. In Figure 33 these parameters are denoted by w, h, and b respectively. The black region is shown to be centered with respect to the cell, but this is not essential.

The calibration band is printed in such manner that the cell width w is approximately equal to the lenticular pitch p. In order to have the black region resolved in the print, its width must be larger than the thinnest line which the printer can print. The width of such line can be estimated as explained in paragraph 10.2 below.

A scheme of the whole calibration band image is shown in Figure 34. The black regions may be regarded as a grid having a pitch w.

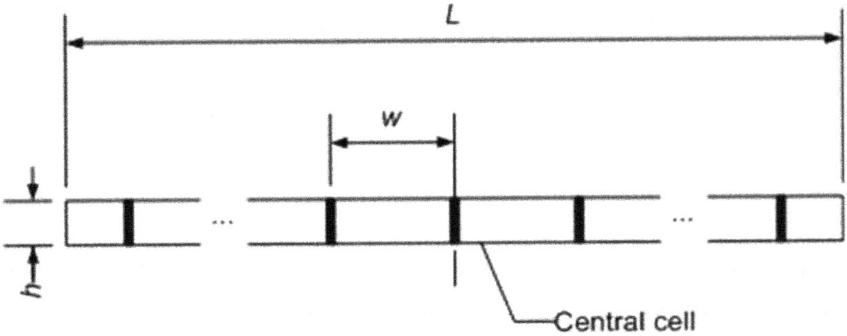

Figure 34: A calibration band image.

In Figure 34 it was assumed that the number of cells is uneven, so there is a central cell. This is not essential, but is convenient for the subsequent discussion. The number of cells in the band image by N_c, the band length L and the cell width w are related as follows:

$$L = N_c w \ .$$

5.2.3 *The calibration pattern image*

A calibration pattern image is a vertical stack of calibration bands images such that the cell width of each band is increased by an increment Δw with respect to the one above it. Usually, an increment much smaller than the cell width is used.

The bands are stacked so that their centers coincide with the center of the pattern. A scheme with such a pattern is shown in Figure 35. The increment Δw in the cell size creates an increment ΔL in the band length:

$$\Delta L = N_c \Delta w$$

It will be further assumed that the number of bands N_b in the pattern is uneven. In this case, there is a central band. The bands can be denoted by an index k, which takes values from $-(N_b\text{-}1)/2$ to $(N_b\text{-}1)/2$. The index $k = 0$ corresponds to the central band. The cell width of the central band will be denoted by w. Correspondingly, the cell width of the k-th band is given by

$$w_k = w + k\Delta w \ . \qquad \textit{Equation 29}$$

The vertical gap between successive bands is not critical. It is typically chosen as a fraction of the band height *h*.

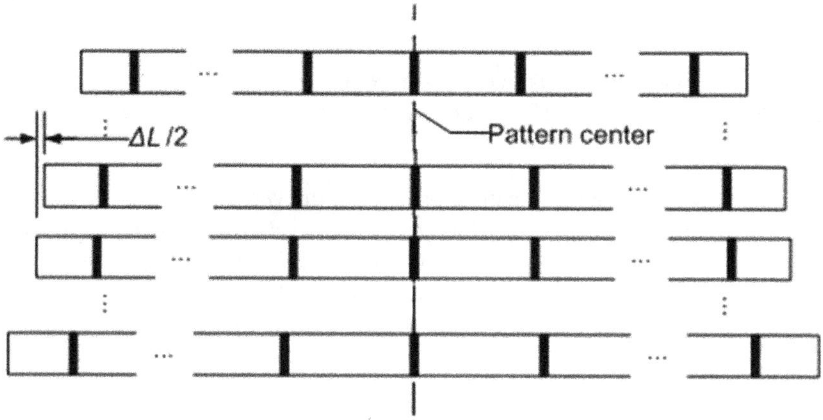

Figure 35: A centered calibration pattern image.

Note that the sizes of the calibration pattern images are in arbitrary units. They acquire real length units only after the pattern is printed. The printed sizes of the various pattern objects will be denoted by the same symbol as used to denote them in the image with a prime added. Thus, for instance, w' and L' will denote the printed size of w and L respectively. The ratios between sizes in the same direction (horizontal or vertical) are conserved in print. Thus, for instance

$$\frac{w}{L} = \frac{w'}{L'} \ .$$

5.3 Printing the calibration pattern image

A certain pixel size b_p must be assigned to the black region width to rasterize the calibration pattern image. As discussed in paragraph 3.4, it is recommended that $b_p \geq 2$ to reduce the rasterization blur. The pixel size w_p of the central band cell width will be

$$w_p = b_p \frac{w}{b} = b_p Q \ , \qquad \textit{Equation 30}$$

where Q denotes the ratio w/b.

The printing requirement of the calibration pattern image is that the cell width of the center band in the printed picture will be equal to a particular pre-determined value w'. A preferred printing density must be used for the horizontal direction. Denoting the chosen horizontal printing density by ρ_h, the following relation holds:

$$w_p = b_p Q = \rho_h w' \,, \qquad \text{Equation 31}$$

from which it follows that

$$b_p = \frac{\rho_h w'}{Q} \,. \qquad \text{Equation 32}$$

For example, suppose one wants to calibrate a 60lpi sheet and sets $w' = 1/60$ inch. Using a 720dpi printer, $b_p = 12/Q$ according to Equation 32. One can choose $Q \leq 6$ to get $b_p \geq 2$.

As in the case of interlaced pictures, printing in the vertical direction does not need to be done with high accuracy. Therefore, the vertical printing dot density ρ_v does not need to be a preferred printing density of the printer. It can be chosen so that the printed band height will be equal to a predetermined value h':

$$\rho_v = \frac{1}{h'} \frac{w_p h}{w} \,.$$

5.4 Measuring the optical grid pitch

The calibration picture is obtained by laying the calibration pattern on the sheet backplane. The pattern must be aligned and centered with respect to the sheet. The optical grid pitch is measured by inspecting the calibration picture. A photograph of such picture from its central normal is shown in Figure 36[1].

[1] Courtesy of Pop3DArt

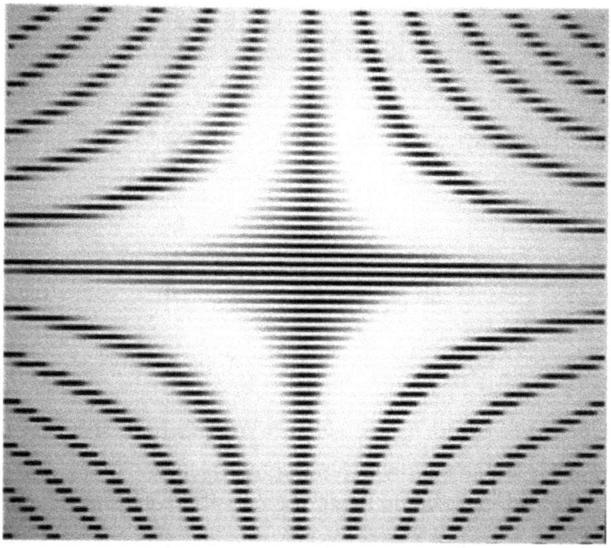

Figure 36: Photograph of a calibration picture taken from its central normal.

The calibration picture pattern exhibits a cross and a set of curves. To understand the origin of the vertical branch of the cross, let us consider the scheme shown in Figure 37. Since both the pattern and the photograph in Figure 36 are centered, the central optical gridlines fall on black regions in all bands, as illustrated in Figure 37. For this reason, all bands centers in Figure 36 are black. This creates the vertical branch of the cross.

A band for which all optical gridlines fall on black regions will appear black. Such bands will be called "matched." In general, there will be more than one matched band in the calibration picture. For example, in Figure 36 there are two black bands. These bands constitute the horizontal branch of the cross.

The indices of the matched bands depend on the viewing distance. When the viewing distance increases, the optical grid pitch decreases, and the position of the matched bands moves accordingly.

The value of the optical grid pitch is estimated by taking the average of the cell widths of all matched bands in the pattern picture. In the case shown in Figure 36, it is the average of the cell widths of the two matched bands.

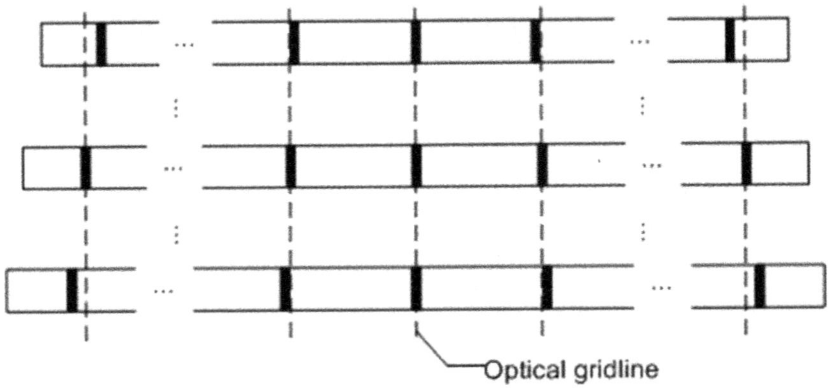

Figure 37: The band calibration pattern with optical gridlines.

The curves seen in Figure 36 follow a hyperbolic trajectory. They reflect the locations in which the optical gridlines fall on black regions of the calibration bands.

Many commercial interlacing programs require the grid density ρ (number of grid cells per unit length) rather than the cell width value. These quantities are reciprocals of each other.

5.5 The number of matched bands

A calibration picture should show at least one matched band. This can be guaranteed by choosing the band length increment ΔL to be smaller than $2b$. Successive bands with a length increment equal to $2b$ are shown in Figure 38 and Figure 39. The cell sizes are slightly different in each band, but the difference is too small to be noticed in the three central cells. However, as the edge is approached, the mismatch builds up, and a visible discrepancy appears between the black region center and the corresponding optical gridline.

Figure 38 shows a typical case in which the two extreme optical gridlines are contained in the black regions of the matched band shown by broken lines. In this case, there is only one matched band. Figure 39 shows a singular case in which the two extreme optical gridlines fall precisely on the boundaries of the black zones of two adjacent bands. In this case, there are two matched bands, and they are also shown in green. We invite the reader to convince themselves that, except the singular case shown in Figure 39, when $2b = \Delta L$ there will always be a single matched band.

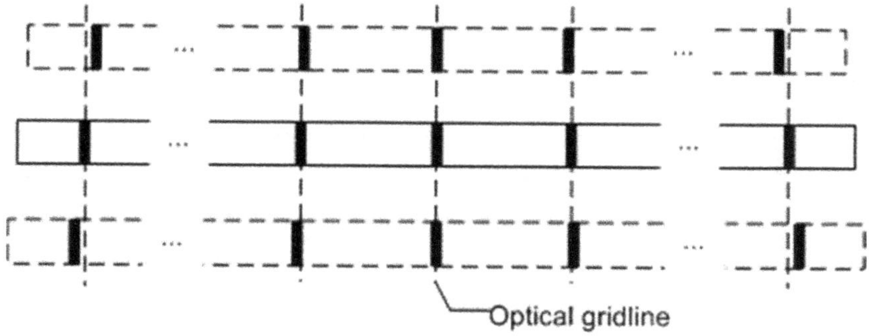

Figure 38: A band calibration pattern with 2b = ΔL displaying a single matched band (solid line).

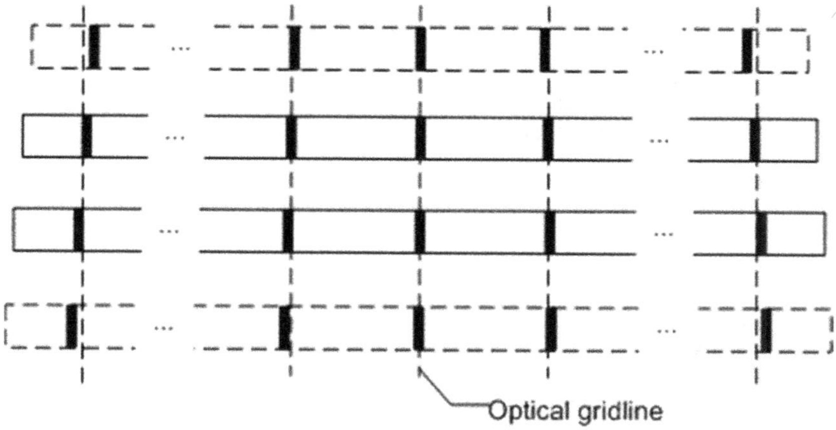

Figure 39: A band calibration pattern with 2b = ΔL displaying two matched bands (solid line).

Choosing a length increment smaller than $2b$ will result in general in more than one matched band which will appear black in the calibration picture pattern. The number N_m of matched bands for a general line increment ΔL is

$$N_m \approx \frac{2b}{\Delta L} = \frac{2}{N_c Q}\frac{w}{\Delta w} \quad . \qquad \text{Equation 33}$$

5.6 The measurement accuracy

Figure 40 shows three special matched bands. The central band represents a perfect match between the cell width and the optical grid pitch. The length of this band was denoted by L'. The lengths of the top and bottom bands are $L' - b'$ and $L' + b'$ respectively. Any band in a given calibration pattern whose length is in the interval $[L'-b', L'+b']$ will appear matched.

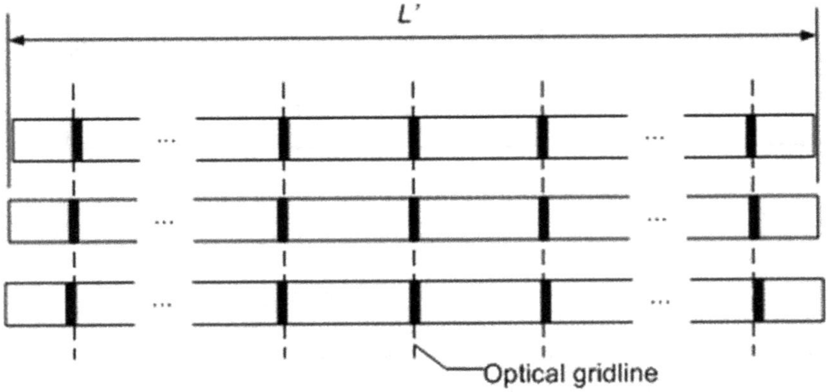

Figure 40: The three extreme matched bands marking the uncertainty interval.

Let us denote by ε the cell increment corresponding to the three bands of Figure 40. Their lengths can be expressed as follows:

$$L' - b' = N_c(p_o - \varepsilon),$$

$$L' = N_c p_o,$$

$$L' + b' = N_c(p_o + \varepsilon),$$

from which it follows that

$$\varepsilon = \frac{b'}{N_c} = \frac{w'}{QN_c}\ . \qquad \textit{Equation 34}$$

It may be concluded that for any given calibration pattern, there is a certain range of cell sizes which will appear matched. The size of this range is 2ε, and it is called "uncertainty interval." The relative size of the uncertainty interval is given by

$$\frac{2\varepsilon}{p_o} = \frac{2b'}{N_c p_o} \approx \frac{2}{N_c Q} = N_m \frac{\Delta w'}{w'} \qquad \textit{Equation 35}$$

If the edges of the uncertainty interval are precisely known, the optical grid pitch could be determined with zero error. However, the actual measurement error depends on the number of matched bands shown in the calibration picture. These bands can be regarded as (equidistant) samples of the uncertainty interval. An example of three matched bands is shown in Figure 41. The maximal cell increment, in this case, is $2\varepsilon/3$, and this increment is used in Figure 41.

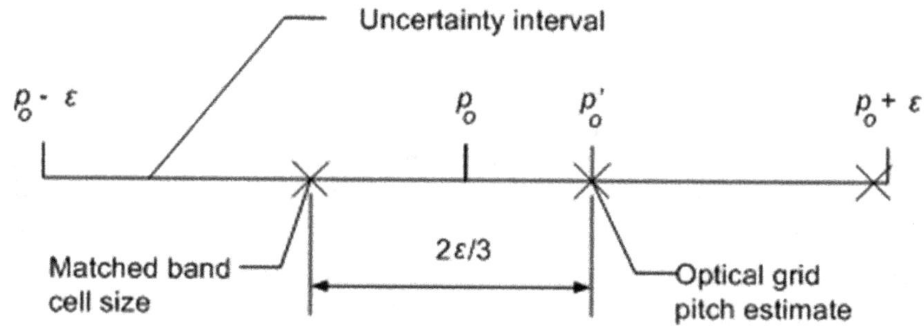

Figure 41: The uncertainty interval with three matched bands. The crosses denote matched bands cell sizes.

According to the measurement procedure, the optical grid pitch is the average of the cell sizes of all matched bands, which, in this case will be the cell size of the middle band. Denoting this estimate by p'_o, it can be seen that the maximal error in the optical grid pitch will be

$$\left|p'_o - p_o\right|_{Max} = \frac{\varepsilon}{3} .$$

This error is called the "sampling error." It is easy to prove that the maximal sampling error for the case of N_m matched bands is ε/N_m.

The sampling error is not the only error mechanism in this measurement. The uncertainty interval boundaries are not sharply defined; for a band with a cell size near the boundary, there is uncertainty in judging whether it is matched or not. It can be said that there is a certain transition region around the uncertainty interval edges in which there is ambiguity concerning matching. This ambiguity contributes an additional error in the measured pitch size which is called the "transition region error." Denoting this width by δ, the maximal measurement error can be written as

$$\text{Maximal error} \approx \frac{\varepsilon}{N_m} + \delta . \qquad \textit{Equation 36}$$

According to Equation 36, for $N_m >> \varepsilon/\delta$ the error is dominated by the transition region.

One of the mechanisms contributing to δ is the rasterization blur, but in general, there are others. Experience shows that δ depends on the black region width b'. When b' is close to the printer resolution, the contrast in the calibration picture is reduced, making it more difficult to judge whether a given band is matched or not. When b' is considerably larger than the printer resolution, one may expect δ to be significantly smaller than ε, so that for small (typical) values of N_m it can be neglected in Equation 36.

5.7 Choosing the calibration pattern parameters

5.7.1 The central band cell width

A given calibration pattern will cover only a rather small range of cell widths. It often happens that none of the bands in the pattern appears black. In such a case, the pattern must be printed again with another range of cells widths. To avoid this, one should choose the best estimate of the optical grid pitch as the cell width of the central band.

Many manufacturers quote the thickness and the lenticule density of the lenticular sheet. The refraction index is usually not quoted, but the value of 1.6 is usually a reasonable estimate. With this data, it is possible to estimate the optical grid pitch at any given distance using Equation 12. For instance, let us assume that the lenticular sheet density and thickness are 41.2lpi and 2mm respectively. Then, for viewing distance of 500mm, one should choose

$$w' = \frac{25.4}{41.2}\left(1 + \frac{2}{1.6 \cdot 500}\right) = 0.6180\,\text{mm}$$

for the central band cell size.

If the exact value of the density is not quoted, one should use the specified lenticule density of the sheet as a first guess. Since this is an approximation, there is no need to make the viewing distance correction. So, for a 40lpi sheet, one should choose

$$w' = \frac{25.4}{40} = 0.635\,\text{mm}$$

for the central band cell size.

5.7.2 Choice of the number of matched bands

Ideally, the chosen value of N_m should be sufficiently large so that the sampling error will be small compared to the transition region error, namely, much larger than ε/δ. On the other hand, choosing a small value of N_m has the advantage of maximizing the optical grid pitch range which is covered by the pattern, so the actual choice is a compromise between these two considerations. Since the value of δ is unknown, N_m is usually chosen arbitrarily to be between 1 and 5.

5.7.3 Choice of the cell size increment

The choice of $\Delta w'$ is determined by the relative accuracy requirement. The maximal measurement error is given by Equation 36. However, since δ is unknown, there is no rigorous estimate of the measurement error. For the sake of this discussion, let us assume that the transition region contribution to the error is negligible. In this case, the relative accuracy ζ is given by

$$\zeta = \frac{\varepsilon}{N_m p_o} \approx \frac{\Delta w'}{w'}, \qquad \text{Equation 37}$$

where in the second step Equation 35 was used. The value of ζ is determined by the required accuracy of the calibrated viewing distance, as discussed in 5.1. Thus, the relative increment in the cell size can be computed from Equation 37.

5.7.4 *Number of bands*

A band calibration pattern covers a certain range of cell widths. This range is determined by the increment $\Delta w'$ and the number of bands. The cell width values interval I_p is

$$I_p = \left[w' - \frac{(N_b - 1)\Delta w'}{2}, w' + \frac{(N_b - 1)\Delta w'}{2} \right].$$

If the pitch of the optical grid falls outside this interval, there will be no matched band in the picture, and one must print the pattern again with a different interval.

There is a practical limit to the number of bands which can be printed. To minimize the number of attempts, it is common to print the first pattern with a large value of $\Delta w'$, say, ten times the value of the accuracy requirement. This pattern yields an approximate value of the optical grid pitch. Next, the band calibration pattern with the required value of $\Delta w'$ is printed using the approximate pitch as the cell size of the central band.

5.7.5 *The Q parameter*

Once the value of N_m is chosen, the value of ζ also determines the value of ε. For a given w, the value of ε determines the value of the product $N_c Q$ by Equation 34.

It is advantageous to choose a value of Q as small as possible, to maximize the width of the black regions. Thicker black regions yield calibration pictures with better contrast, making it easier to determine which band is matched, thus reducing δ. However, for a given accuracy requirement, a smaller value of Q requires a larger value of N_c (Equation 35), or, in other words, a wider pattern.

5.7.6 *Example*

Table 10 shows an example of parameter values of a band calibration pattern. There are three categories of parameters: given, chosen, and calculated. The parameters N_b and h were not included in the table, as these are chosen purely by convenience, and have no impact on the measurement accuracy.

The value of b' in this example is 150μm, which is about three times the printer resolution (paragraph 10.2). Therefore, this pattern picture will possess a good contrast.

Table 10: Parameters for band calibration pattern: an example.

Type	Symbol	Description	Value
Given	p	Lenticular pitch	0.6350mm
	t	Lenticular sheet thickness	2mm
	n	Index of refraction	1.6
	L'	Width of the printed central band	500mm
	ς	Required relative accuracy	10^{-4}
	V	Viewing distance	1000mm
Chosen	N_m	Number of matched bands	3
Calculated	N_c	Number of cells, L'/p	~787
	w'	Central band cell printed size (Equation 12)	0.6358mm
	$\Delta w'$	Printed cell increment (Equation 37)	0.127μm
	Q	Cell width to black region width ratio (Equation 33)	4.24
	b'	Printed black region width	150μm

5.8 Using an interlacing algorithm to create the central band

It is quite common to create the central band by interlacing a sequence with the same width and length as the planned lenticular picture. The resulting band width is the same as the interlaced image of the lenticular picture, thus using the maximal width allowed by the existing lenticular sheet. A special sequence is used to create the band image: all sequence images are either black or white, and images of the same color are grouped. By denoting the sequence length and the number of black images by N and n_b respectively, one gets

$$Q = \frac{N}{n_b} \,.$$

Since N is given and n_b must be an integer, this limits the choice of the values of Q which can be used with this method, but this limitation is irrelevant for all practical purposes.

5.9 The lenticular pitch

The calibration procedure that has been described is used to measure the optical grid pitch, but not the lenticular pitch p. To make a single lenticular picture, it is not necessary to know p. However, it is useful to be able to determine the optical grid pitch for a given viewing distance. If the parameters p, t, and n are known, then Equation 12 can be used to calculate p_o for any viewing distance V. The knowledge of p is also needed for the layout of several 3D pictures on a single lenticular sheet.

The lenticular sheet pitch can be measured directly from the calibration pattern picture by observing it from a large distance. The relative discrepancy between the optical and the sheet pitch values is derived from Equation 12:

$$\frac{p_o - p}{p} = \frac{t}{nV} .$$

This means that for an adequate estimation the ratio t/nV should be of the order of 10^{-4}. For thick lenses, this requires an inspection from a distance of tens of meters, and may be inconvenient.

Alternatively, the value of p can be derived from several calibration measurements at different distances by utilizing standard statistical methods. Suppose that a set of optical grid pitch values p_o^i was measured for a set of viewing distances V_i. Equation 12 can be rewritten as

$$p_o = pz , \qquad \textit{Equation 38}$$

where

$$z = 1 + \frac{t}{nV} .$$

The value of p can now be derived by using the set of measured points $[p_o^i, z_i]$ and Equation 38 as inputs to a standard linear regression analysis. To compute z, one needs to know the value of the ratio t/n. The thickness t can be either measured or taken from the lens sheet data. If the value of n is unknown, the value 1.6 can be used. Alternatively, it is possible to determine the ratio t/n by seeking the value which gives the best regression fit.

Figure 42 shows an example of such measurement. The chart shows 11 measurements of the optical grid pitch for different distances. All measurements used the same calibration pattern picture. The lens has a specified density of 60lpi.

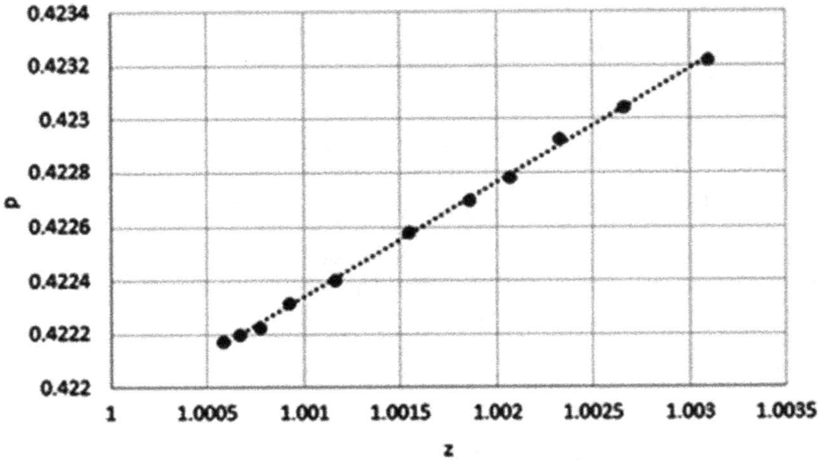

Figure 42: Measurement of the lenticular sheet pitch. The vertical axis is the measured p_o.

The dotted line shows the linear regression result. The calculated value of the pitch for this lenticular sheet was 0.42192mm, which corresponds to 60.201lpi. According to Table 8, the required density for a matched pixel grid is 60.075lpi. Therefore, this lenticular sheet cannot benefit from reduced rasterization blur.

5.10 The continuous calibration pattern

5.10.1 General

The calibration pattern shown in Figure 36 is a discrete version of a certain continuous pattern. The continuous calibration pattern is obtained as a limit in which one holds the top and bottom bands cell sizes constant, decreases the bands height *h* and the gap between them to zero, and increases the number of bands to infinity. Under these conditions, the bands become a continuum of lines.

The continuous calibration pattern can represent a continuum of lines only in its continuous format. Its printed version consists of finite-size pixels. Therefore, it is basically identical to the band pattern, with the only difference being that the bands height size equals the vertical printed pixel size, which is much smaller compared to the band height in the calibration pattern.

The continuous calibration pattern has one obvious advantage: the number of matched "bands" in such pattern can be chosen to be very large, and therefore make the sampling error insignificant. This increases the accuracy of the continuous pattern compared to a band pattern with the same width.

5.10.2 Graphical construction

The continuous pattern is created from a single band image, like the one shown in

Figure 34, except that a much larger value of h is used, as shown in Figure 43, instead of $h << L$, as implied by Figure 34. Again, it is assumed that the number of cells is uneven.

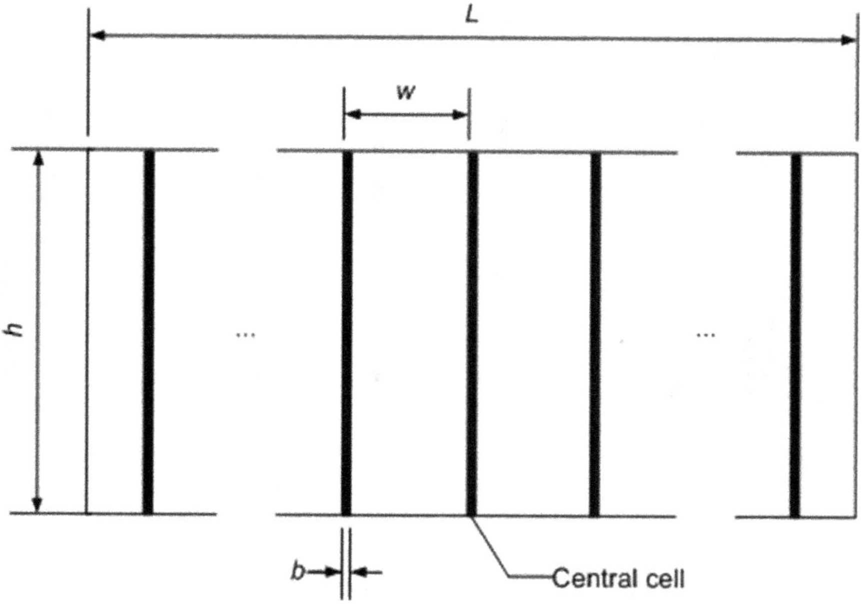

Figure 43: The image used to create the continuous calibration pattern.

Next, this pattern is skewed by moving the top corners towards the center and the bottom corners away from it. All displacements are chosen to be equal. The result of this transformation is shown in Figure 44. Note that because the number of cells is uneven, there is a central cell, and its center line remains vertical after the transformation, although it is converted from a rectangle to a trapezoid. Since the skewing is distributed equally between all cells, the deviation of the central section from rectangularity is tiny, and is not shown in Figure 44.

In this case, the graphical content of the image is not rectangular, and therefore it must be drawn on a rectangular canvas, as explained in 3.8. In this case, the canvas width is $L + 2\eta$ and height h. The exposed canvas areas are painted in Figure 44 in black.

Strictly speaking, the skewing transformation also converts the black regions into trapezoids, so that their width at the top is slightly smaller than in the bottom. This, however, has a negligible effect, and will be disregarded.

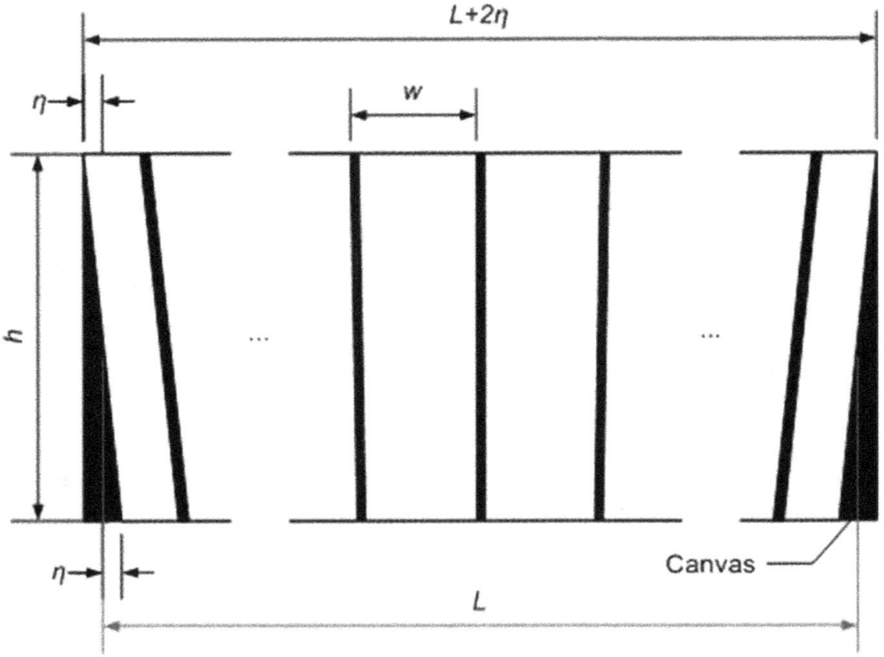

Figure 44: The continuous calibration pattern.

5.10.3 The continuous calibration pattern analytics

The continuous calibration pattern is characterized by the parameters listed in Table 11. These parameters have analogous counterparts in the band calibration pattern. The skew shift η replaces the cell increment Δw.

Table 11: The parameters of the skewed calibration pattern.

Symbol	Description
w	Cell size at half height
N_c	Number of cells
h	Height
η	Skew shift
b	Black region width

In this pattern, the pitch varies continuously along the vertical direction. The pitch or the density at a given position can be determined with the help of Figure 45.

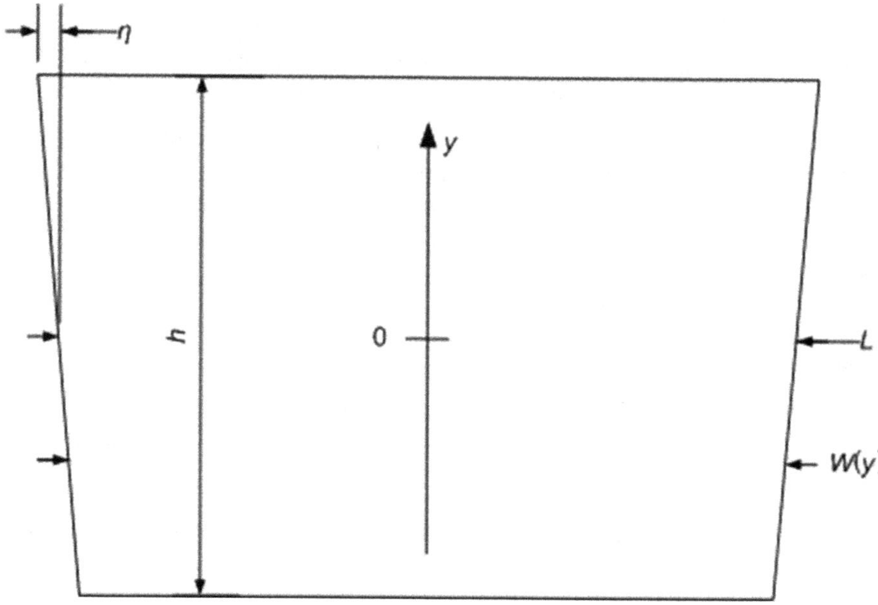

Figure 45: Geometry of the continuous pattern.

Let us define a vertical axis y with an origin at the pattern center. The width $W(y)$ of the pattern varies linearly with y, with the following values at the top, center, and bottom:

$$W(h/2) = L + 2\eta,$$

$$W(0) = L,$$

$$W(-h/2) = L - 2\eta.$$

The pattern width at half height was denoted by L (which is also the original pattern width before applying the skewing distortion). From this it follows that

$$W(y) = L + \frac{4\eta y}{h} \ .$$

The cell size c is a function of y:

$$c(y) = \frac{W(y)}{N_c} = w + \frac{4\eta y}{N_c h} \ . \qquad \textit{Equation 39}$$

The cell density $\rho(y)$ as a function of y is given by

$$\rho(y) = \frac{1}{c(y)} \approx \rho_0\left(1 - \frac{4\eta y}{Lh}\right), \qquad \textit{Equation 40}$$

where $\rho_0 = 1/w$.

5.10.4 The continuous calibration pattern picture

Analogous to the printing procedure of the band calibration pattern, it is required that the printed cell width w at $y = 0$ will have a given value w'.

The line $y = 0$ may be considered as a calibration band with an infinitesimally small height. It represents a black/white periodic grid, with a cell width w and a black region width b, and its Q value is w/b. Now Equation 32 can be used to determine the pixel size of b.

The height of the printed pattern can be adjusted to a convenient size by using an appropriate value of the vertical printing density ρ_y.

Let us rewrite Equation 39 with primed quantities (which correspond to a printed picture of the pattern image):

Equation 41

$$c(y') = w' + \frac{4\eta' y'}{N_c h'} .$$

This equation gives the cell size corresponding to vertical coordinate value y' of the pattern picture. Differentiating $c(y')$ with respect to y' gives

$$\frac{dc}{dy'} = \frac{4\eta'}{N_c h'} = \frac{1}{M} ,$$

where the parameter M was introduced. This parameter represents the vertical displacement in the pattern corresponding to a unit change in the cell size and may be regarded as the pattern magnification.

5.11 An example of calibration using the continuous pattern[2]

When the printed continuous pattern is covered by a suitable lens sheet, one obtains the continuous calibration picture. Figure 46 shows a photograph of such picture taken from 1m. In this picture, a 56cm wide 40lpi sheet was used. The horizontal lines correspond to constant cell widths, the values of three of which are shown in the figure. All annotated values in Figure 46 are in μm. The cell increment between successive horizontal lines is 1.29μm.

---

[2] Courtesy of Pop3DArt

Table 12: Skewed pattern parameters.

Symbol	Description	Value
L'	Printed pattern width at half height	560mm
w'	Printed cell size at half height	613.8μm
N_c	Number of cells	~886
h'	Printed height	400mm
η'	Skew shift	2.8mm
b'	Black region width	150μm
M	Magnification	$3.16 \cdot 10^{4}$

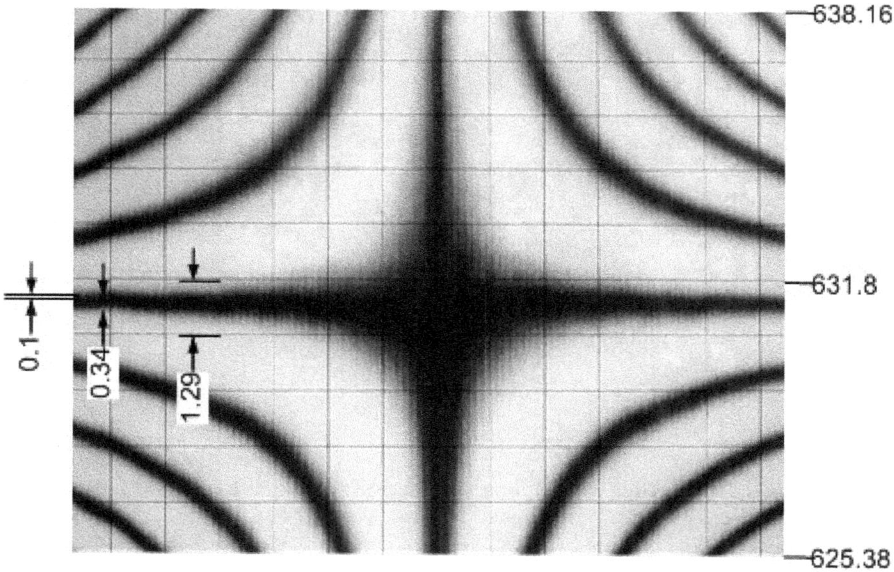

Figure 46: The continuous calibration picture. The numbers on the right represent cell widths in units of μm. Cell increment between horizontal lines is 1.29μm. The other annotated numbers denote the widths of the uncertainty interval and the transition region in units of μm, as inferred from the picture.

Using Equation 34, the value of ε corresponding to Figure 46 is:

$$\varepsilon = \frac{b'}{N_c} = \frac{150\mu m}{886} = 0.17\mu m \ .$$

Thus, the width of the uncertainty interval is 0.34μm. According to the discussion in 5.5, all lines with cell sizes in this interval will be matched, and therefore the height of the horizontal black line at the edge is the uncertainty interval width. This quantity can be measured in Figure 46 using the horizontal lines as a ruler, and it turns out to be 0.34μm, in complete agreement with the analysis.

The transition region width can also be estimated from Figure 46, and it turns out to be approximately 0.1μm. This quantity determines the measurement accuracy with the continuous pattern. Using straightforward methods, it is possible to estimate the optical grid pitch in this case with an accuracy of at least 0.02μm. The corresponding relative accuracy is $3 \cdot 10^{-5}$. This is significantly better than the band calibration example which we considered in Table 10, which had a similar physical width and an identical value of b'.

Finally, let us estimate the required vertical printing density for the pattern in this example. The estimated ratio ε/δ for this pattern is ~2. To make the sampling error negligible, it is necessary to have N_m much larger than 2. Let us estimate the vertical printing density which gives $N_m = 20$.

If the pattern is printed with a vertical density of ρ_v, then the change Δc in the cell size over one vertical pixel is

$$\Delta c' = \frac{1}{M\rho_v} \ .$$

The number of matched bands N_m is given by

$$N_m = \frac{2\varepsilon}{\Delta c'} = \frac{2Mb'\rho_v}{N_c} \ ,$$

and the vertical printing density can be expressed as

$$\rho_v = \frac{N_c N_m}{2M} \frac{1}{b'} \ .$$

Using the values given in Table 12, $N_m = 20$ and expressing b' in inches ($5.9 \cdot 10^{-3}$) gives $\rho_v = 47.5$dpi. Any vertical printing density higher than that will create a pattern with a negligible sampling error. The pattern in Figure 46 was printed with $\rho_v = 80$dpi, so for this pattern $N_m \sim 34$.

Chapter 6
Depth imaging in a 3D lenticular picture

"… all influence is bad, but a good influence is the worst in the world."

- Oscar Wilde, in "A woman of no importance"

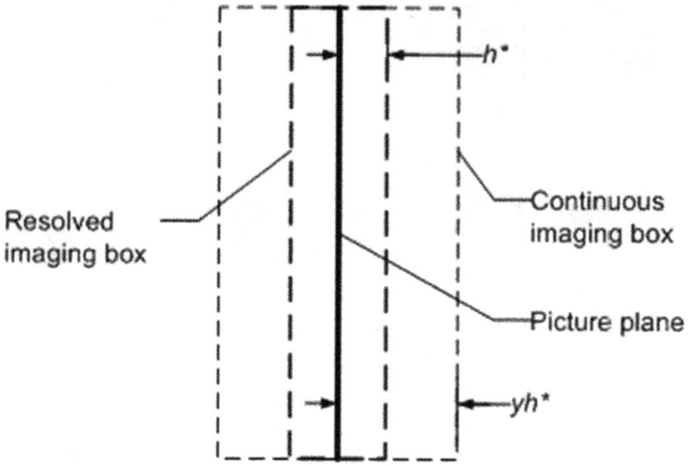

6.1 Real and imaged objects

The discussion of 3D imaging is often obscured by confusion between real and imaged objects. The essence of this confusion is very effectively conveyed by the famous picture made by Rene Magritte, titled "This is not a pipe," shown in Figure 47 (The treachary of images, 1929).

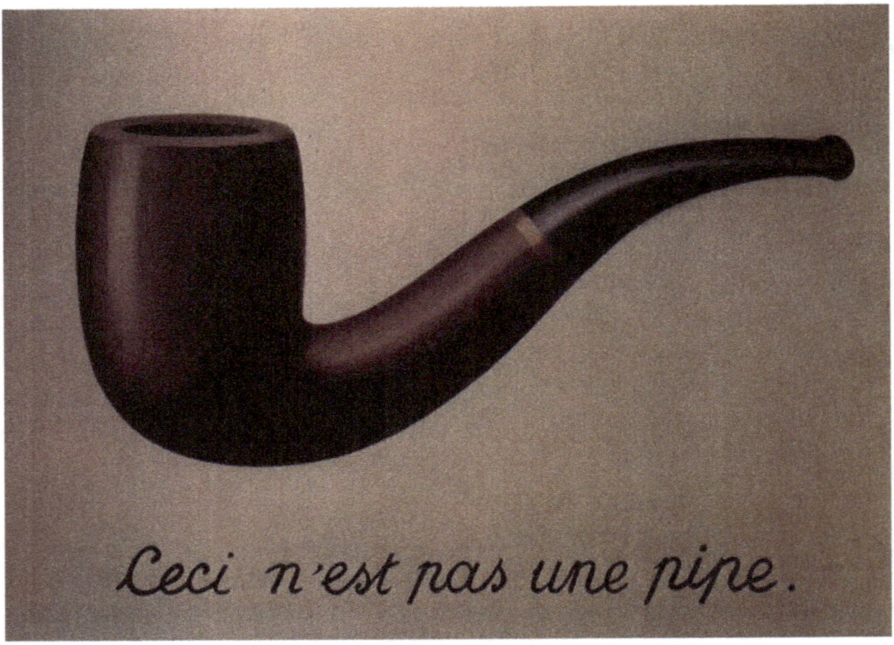

Figure 47: Magritte's "This is not a pipe" picture

A real object is a physical entity with physical properties like dimensions and weight. Its real location is independent of the location of the 3D picture. The imaged object is a virtual entity, and is bonded to the 3D picture which displays it; if the picture moves, it moves together with it. The imaged object can have the same spatial geometry as the real object, or it can be distorted. The most common geometric distortion is scaling. Scaling is very common, to the extent that a display with the original object size may be regarded as a rare exception.

With the advent of 3D computer graphics, it became possible to create digital objects. Such objects are represented by digital files rather than by real physical substance, but they can be imaged like real objects. In the present context, digital objects are also considered to be "real."

Throughout this chapter, imaged objects will be referred to explicitly as "imaged" to avoid this confusion.

6.2 The main viewing interval

The interlaced image deposited at the lenticular sheet backplane is made up of an array of vertical cells having a specific pitch p_p. It is assumed again that the sheet has a central lenticule.

The main viewing interval is a line in the imaging plane and is introduced in Figure 48. The intersection of a printed cell with the imaging plane is an interval which will also be referred to as "cell." The main viewing interval is the image of the central printed cell at a distance V_0 from the axes plane (which, for most practical purposes, may be regarded as the picture plane). The reader may convince themselves that all cells of the interlaced image are imaged by their corresponding lenticules to the main viewing interval. As explained in paragraph 4.4, for any observation point on the main viewing interval, the picture will display one of the sequence images over its whole aperture.

The angle subtended by the main viewing interval with respect to the central axial point is denoted by η. From Figure 48 it follows that

$$\tan\left(\frac{\eta}{2}\right)=\frac{np_p}{2t}=\frac{np}{2t}\left(1+\frac{t}{nV_0}\right),$$ *Equation 42*

where Equation 17 was used in the second step. Since in all practical cases $t << nV_0$, it follows that

$$\tan\left(\frac{\eta}{2}\right)\approx\frac{np}{2t}=\frac{R_p}{2}.$$

Comparing this with Equation 7, it may be concluded that

$$\eta\approx\alpha,$$

where α is the lenticule viewing angle.

When the observer moves their eye from one edge (A) of the main viewing interval to the other (B), they will see all of the picture sequence images displayed in sequence. It is assumed that both of the observer's eyes are on the main viewing interval, each one located at a different point. When the observer looks at the picture with both eyes, each eye will see a different image. If these images display the perspective projections of a three-dimensional object on the picture plane from points which correspond to the location of the eyes, then stereopsis will be invoked and will cause a convincing 3D illusion (Leuder, 2012).

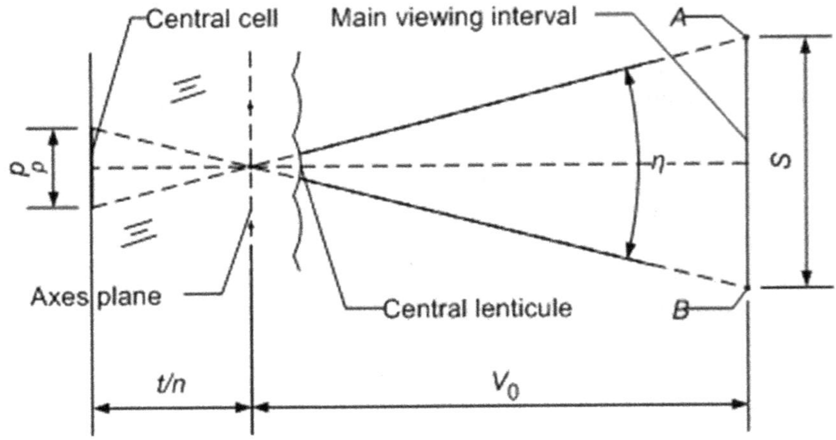

Figure 48: Main viewing interval.

Ideally, different images should be displayed for any two distinct points on the main viewing interval, regardless of how close they are to each other. Real pictures can show only a finite number of different images. This fact limits the depth which can be displayed in a 3D lenticular picture.

6.3 Depth imaging with a 3D lenticular picture

Let us consider an object which is imaged at a certain distance from the picture plane. For this discussion, it is convenient to choose an object which is a thin vertical line, as shown in Figure 49. The object can be imaged either in front of the picture or behind it. An object imaged in front of the picture will be "protruding," and behind the picture "recessing." The distances of protruding and recessing imaged objects from the picture plane will be denoted by h_f and h_b respectively, as shown in Figure 49.

At each observation point on the main viewing interval, the picture should display the corresponding perspective projection of the object on the picture plane. Figure 50 shows the projection geometry for the edges of the main viewing interval. The protruding and recessing imaged objects are shown as black spots. The centers of these objects are mapped by a perspective projection from point A to points A' and A''. Likewise, these centers are mapped by a perspective transformation from point B to points B' and B''.

Figure 50 shows also a general interval Y on the main viewing interval. The interval P_y on the picture plane is obtained by projecting the edges of Y through the center of the protruding imaged object. It can be easily shown that P_y is linear in Y, so that

Equation 43

$$P_y = \frac{Y}{S} P .$$

A similar relation exists for the parallaxes corresponding to the receding object.

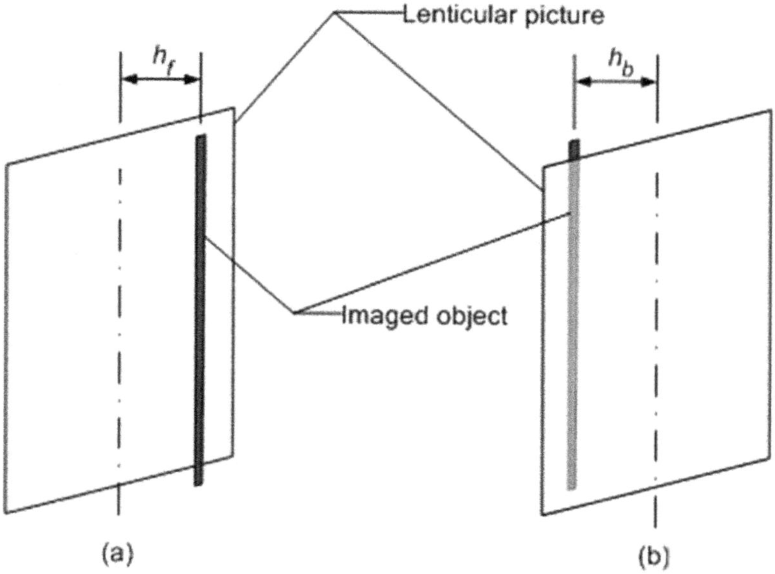

Figure 49: Imaging an object outside the picture plane. (a) Object imaged in front of the picture (protruding), (b) object imaged behind the picture (recessing).

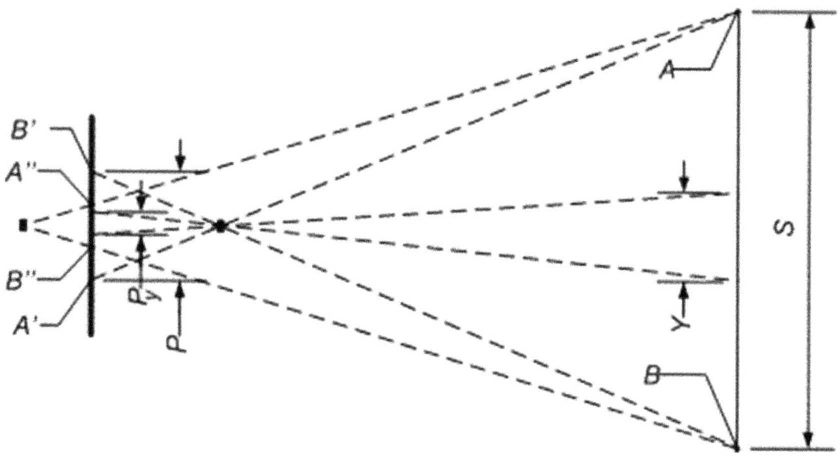

Figure 50: Projections from the edges of the main viewing interval.

Figure 51 shows (schematically) the projections corresponding to the points *A* and *B*. The projections corresponding to the protruding imaged object are shown in the upper part, and those corresponding to the recessing imaged object at the lower part.

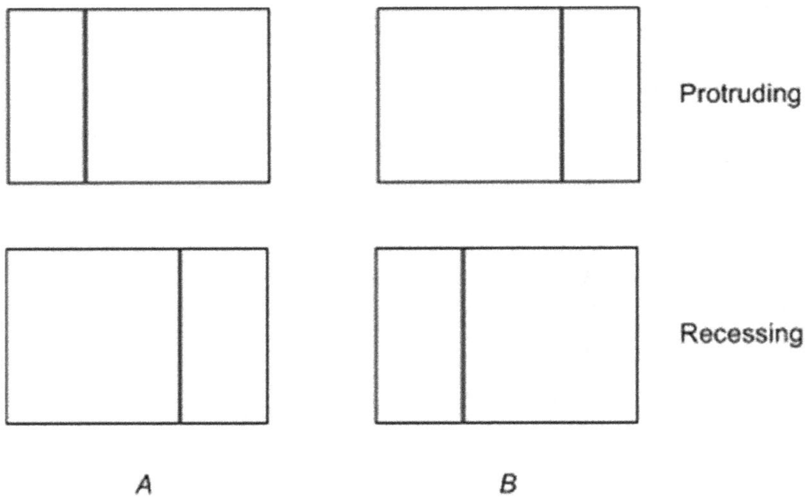

Figure 51: Imaged object projections at the edges of the viewing interval.

When the two projections of one of the objects are superimposed, the image shown in Figure 52 is created. The distance *P* between the two superimposed projections is called the "total parallax." The total parallaxes corresponding to protruding and receding objects will differ in general, even if $h_f = h_b$.

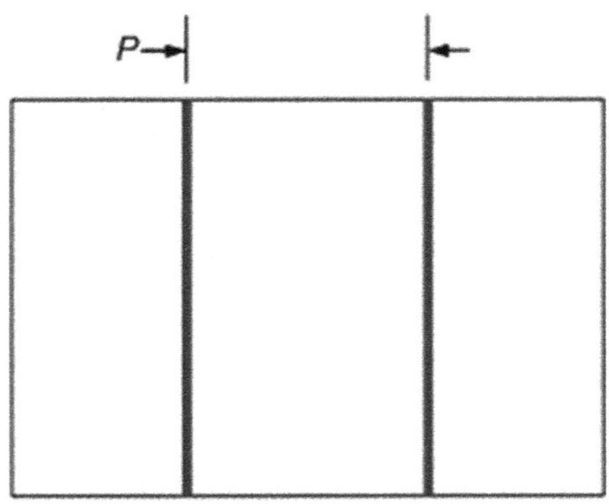

Figure 52: Superposition of the extreme projections.

The line object considered in this example looks the same from all elevations. This is not true for a general object; the projection of a general object depends on both the horizontal and the vertical viewing angles. The image displayed by 3D lenticular pictures does not change when the observer moves their eye in the vertical direction, which is parallel to the lenticules axes. Therefore, 3D lenticular pictures cannot display changes in the object projection in the vertical direction.

Assuming that the imaged object projections correspond to points on the main viewing interval, the observer will see the object correctly only from this interval. Observing the picture from above or below the main viewing interval will introduce geometrical distortions, which will increase with the deviation from the main viewing interval.

In this example, the projections of the object from any two given points on the main viewing interval are identical except for a horizontal shift in position. Any two projections can be superimposed in the manner shown in Figure 52, and the parallax between them can be measured. This parallax is linear in the separation between the two points. There will be two special parallaxes in this discussion:

1. Total parallax, corresponding to the two extreme points of the main viewing interval, and
2. Viewing parallax, corresponding to two points separated by the inter-ocular distance.

In all practical viewing scenarios, the main viewing interval *AB* is larger than the interocular distance. Therefore, the viewing parallax is smaller than the total parallax.

6.4 The relation between imaged object position and its parallax

Using triangular similarity relations, from Figure 48 it follows that

$$S = \frac{nV_0 p_p}{t} \quad \text{Equation 44}$$

As explained in paragraph 4.3, the printed grid pitch p_p must be equal to the optical grid pitch, which, in the simplified model equals the axial grid pitch p_a. Using Equation 6 and Equation 17, *S* can be written as

$$S = R_p V_0 \left(1 + \frac{t}{nV_0}\right) \quad \text{Equation 45}$$

Let *P* be the total parallax of an imaged object (either protruding or recessing). Using Equation 43 with $Y = e$ (where *e* is the inter-ocular distance), the following expression for the viewing parallax P_v is obtained:

$$P_v = \frac{e}{S} P = \frac{eP}{R_p V_0 \left(1 + \frac{t}{nV_0}\right)} \quad \text{Equation 46}$$

In all practical cases, $t << V_0$, so this expression can be approximated by

$$P_v = \frac{eP}{R_p V_0}$$ Equation 47

This approximation will be used from now on.

The relation between the distance of a protruding image object from the picture and its parallax can be computed with the help of Figure 53.

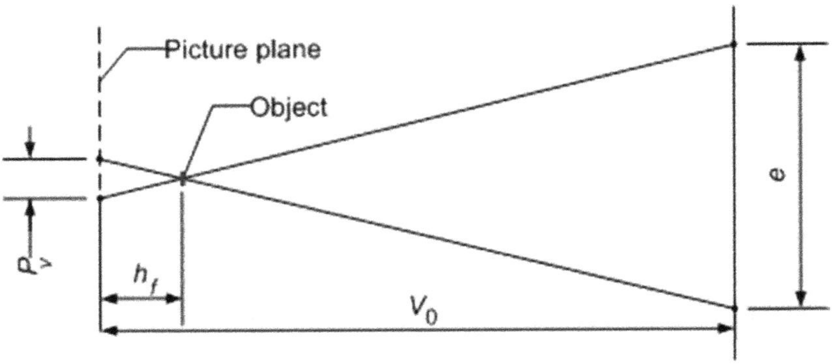

Figure 53: Parallax of an object in front of the lens.

From triangular similarity, it follows

$$\frac{h_f}{P_v} = \frac{V_0 - h_f}{e} ,$$ Equation 48

so that

$$h_f = \frac{V_0 \cdot P_v}{e + P_v} .$$ Equation 49

A similar analysis for a recessing imaged object reveals that

$$h_b = \frac{V_0 \cdot P_v}{e - P_v} .$$ Equation 50

It is convenient to introduce a new quantity h:

$$h = \frac{V_0}{e} P_v = \frac{P}{R_p} .$$ Equation 51

The distances h_f and h_b can be expressed in terms of h as follows:

$$h_f = \frac{h}{1 + h/V_0} ,$$ Equation 52

$$h_b = \frac{h}{1 - h / V_0} \cdot \qquad \textit{Equation 53}$$

Equation 52 and Equation 53 show the relation between the total parallax P and the distances of the imaged objects from the picture plane.

In most practical cases $h << V_0$, so that the term h/V_0 can be neglected with respect to 1. This is called the “small parallax approximation." Within this approximation, Equation 52 and Equation 53 simplify to

$$h_f = h_b = h, \qquad \textit{Equation 54}$$

so that h becomes the distance of an imaged object from the picture plane for a given total parallax P, regardless of whether it is recessing or protruding. From Equation 54 it follows that within the small parallax approximation, the imaged distances do not depend on the calibration distance, which, in this discussion, is also the viewing distance. These distances are uniquely determined by the total parallax P.

Equation 51 implies that the imaged object distance from the picture plane is inversely proportional to the sheet relative pitch. For this reason, lenses with a small value of R_p are used for 3D applications, in order to increase the displayed depth for a given parallax. A typical value of R_p for 3D applications is 0.5 so that $h \sim 2P$.

6.5 Continuous lenticular pictures

As explained in paragraph 6.1, the picture will display one of the sequence images for any given observation point on the main viewing interval. As the observer moves their eye on the main viewing interval, they may see abrupt changes in the displayed image as one image from the sequence is replaced by another. If the difference between the images is smaller than a certain value, the transitions between them may become imperceptible, and the observer will perceive a continuous evolution in the displayed image as if the picture sequence was infinite. In 3D lenticular pictures, such continuity creates a very convincing illusion of observing a real 3D object. Correspondingly, pictures displaying continuous transition between adjacent images are called "continuous". Pictures which display discontinuous transitions between adjacent images will be called “discrete”.

The object considered in paragraphs 6.3 and 6.4 had the form of a thin vertical line. This object was convenient for the discussion in these paragraphs because it defines a unique parallax. Real objects will have a range of parallaxes since the parallax associated with a given object point depends on its deviation from the picture plane. Therefore, parallaxes can be uniquely associated only with imaged object points, not with the imaged objects themselves.

The translations of a given object point in 3D sequence images are horizontal because they are all projections from points located on a horizontal line. There may be exceptions to this rule, most notably, when the scene shows refracted or reflected objects from curved surfaces. However, these exceptions will be disregarded here, and the parallax will always be horizontal.

As explained in paragraph 2.8.3, the effective resolution element of the lenticular picture in the horizontal direction is the lenticule pitch p. If the change between adjacent images of the sequence is less than the resolution element of the picture, the transition between them will become unnoticeable.

For a given object point, there is a certain parallax P_S between successive images. This parallax is called "incremental parallax," and it is given by

$$P_S = \frac{P}{N} ,$$

where N is the sequence length. In the small parallax approximation, the imaged object point distance h from the picture plane can be expressed in terms of the incremental parallax as follows

$$h = \frac{NP_S}{R_p} = \frac{Nt}{n}\frac{P_S}{p} . \qquad \textit{Equation 55}$$

Picture continuity is achieved if $P_S \le p$ for all displayed object points. Thus, the largest distance h' from the picture plane in which imaged object points are continuously displayed is obtained by substituting p for P_S in Equation 55:

$$h' = \frac{Nt}{n} . \qquad \textit{Equation 56}$$

The quantity h' denotes a boundary for continuous imaging. Any imaged object point whose distance h from the picture plane is smaller than h' will be displayed continuously. Alternatively, it may be said that any imaged point at a distance h from the picture plane will be continuously displayed if the value of N is increased until it satisfies the inequality below:

$$N \ge \frac{nh}{t} . \qquad \textit{Equation 57}$$

The distance h' depends both on the properties of the lenticular sheet (n and t), and the sequence length N.

6.6 The number of resolvable images

The picture resolution sets a limit to the number of images that the picture can display distinctively. When this limit is exceeded, adjacent images in the sequence will necessarily blend into each other, causing blurring. This limit will be called "The number of resolvable images" and will be denoted by N_r.

The number of resolvable images is a primary figure of merit of the technology used to make the lenticular picture. It is determined by the fundamental resolutions of the lenticule optics and the printing technology, as well as by the magnitude of the rasterization blur. The optics and printing resolutions cannot be affected by image processing. The rasterization blur may be minimized if a nearly-matched pixel grid is used (paragraph 3.4).

The number of vertical lines that the printer can resolve over the lenticule width will be considered in Chapter 10. The number of vertical lines that a lenticule can resolve over its width will be evaluated in Chapter 11. In the absence of rasterization blur, the smaller of these two numbers gives a fair estimate of N_r. However, since this quantity plays a fundamental role in 3D lenticular imaging, it is worthwhile to measure it directly.

A practical and straightforward way to measure N_r directly is to use a sequence corresponding to an imaged object in the form of a vertical line positioned outside the picture plane, as illustrated in Figure 49. In this sequence, both the object width and P_s are chosen to be significantly larger than the lenticular pitch p. These choices make the display strongly discretized, and allow individual sequence images in the lenticular picture to be easily discerned.

If the picture can resolve this sequence, one should be able to see a single line (with a single eye). If one is not able to see a single line, then the sequence is not resolved. The number of resolvable images is determined by making pictures of such sequences with increasing sequence lengths, starting with a number which is resolved and finishing with a number which is not resolved.

Let us show an example of such a measurement. A 60lpi sheet with the characteristics shown in Table 13 was used in this example. The optical grid pitch of this sheet matched the printer pitch with an accuracy high enough that a matched grid could be used, and rasterization blur avoided.

Table 13: Parameters of the lenticular sheet used for the measurement of N_r.

	Property	**Symbol**	**Value**	**Units**
Given	Lenticule density	ρ_L	60.20	1/inch
	Thickness	t	1.3	mm
	Index of refraction	n	1.56	-
Calculated	Pitch	p	0.4219	mm
	Relative pitch	R_p	0.5	-
	Viewing angle	α	28°	Degrees
	Acceptance angle	β	48°	Degrees

A pattern was created which contained five horizontal bands and was printed at 2880dpi pixel density, each band corresponding to a different sequence length. The sequence lengths that were used were 6, 8, 10, 12, and 16. All bands have a matched pixel grid except $N = 10$, which is only matched periodically. The pattern characteristics are given in Table 14.

Table 14: Pattern parameters

Property	Symbol	Value	Units
Incremental parallax	P_S	7	mm
Object width	-	3.5	mm
Printing pixel density	-	2880	1/inch
Number of images	N	6, 8, 10, 12, 16	-
Object position	-	In front of picture	-
Viewing distance	-	~30	cm

A constant incremental parallax P_S of 7mm was used for all bands. The object was imaged in front of the lens.

A scan of the printed interlaced image is shown in Figure 54. It consists of very fine printed lines, which are barely resolved in this figure. These lines are shown in the magnified view of Figure 55. The gaps between adjacent lines are nearly equal to the lenticular pitch p.

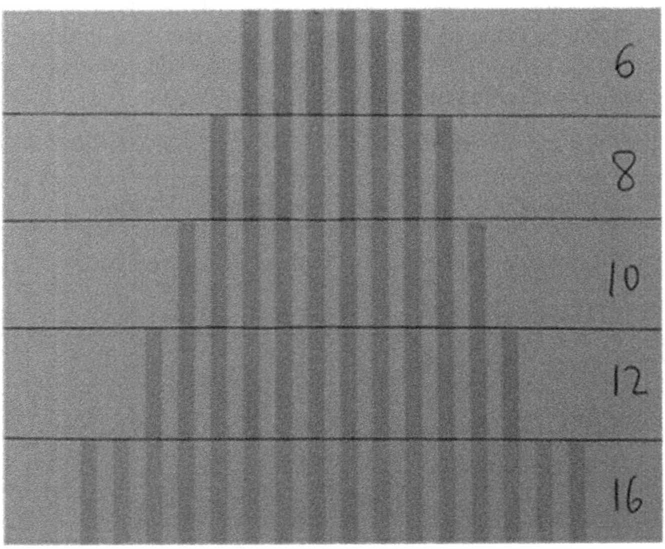

Figure 54: A scan of the pattern used for the measurement of the number of resolvable images.

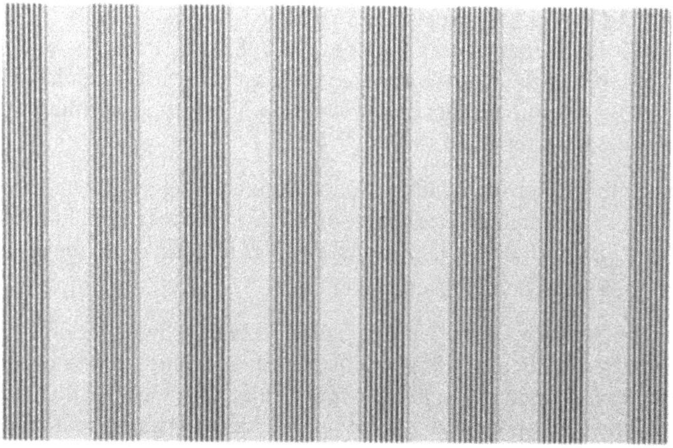

Figure 55: Magnified view of an excerpt from the pattern of Figure 54.

When the interlaced picture shown in Figure 54 is covered by the lenticular sheet and aligned, a lenticular picture is obtained. A photograph from the center of the main viewing interval is shown in Figure 56. This photograph shows one of the center bars.

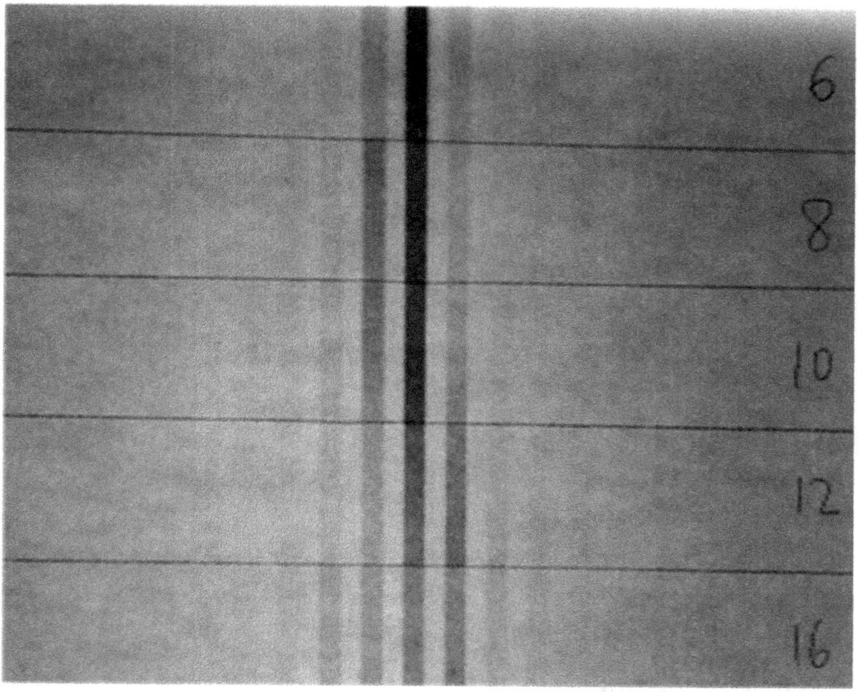

Figure 56: Determination of the number of resolvable images: center view.

Inspection of this photograph reveals that in all cases it is possible to see more than one bar. However, in the bands corresponding to 6, 8, and 10 images one bar stands out as much darker than the rest. This visual contrast serves as a resolution criterion. If the picture displays a group of bars in which one of them appears much darker than the rest, it can be deemed as resolved.

According to this criterion, bands 6, 8, and 10 are resolved. In bands 12 and 16 the contrast is reduced, and the bars are not resolved. So, one may conclude that in this experiment N_r is either 10 or 11. Unfortunately, it is impossible to make this pattern for $N = 11$ without introducing rasterization blur.

The optical analysis suggests that at the picture center, the lenticules can resolve hundreds of lines per lenticule (Chapter 11). However, a study of the printer resolution (paragraph 10.2) shows that it can resolve only about 450 lines per inch, which, in the case of a 60lpi lens corresponds to 7.5 lines per lenticule. The present measurement of the number of resolvable images confirms that the printing technology is the limiting factor in the number of resolvable images.

If the mechanism which limits the resolution is printing, then N_r can be estimated for lenticular sheets with lenticule densities other than 60lpi by using the plausible assumption that N_r is proportional to p, so inversely proportional to the lenticule density. Assuming that in this example $N_r = 10$, it is possible to estimate the number of resolvable images for other lenticule densities by

$$N_r = \frac{600}{\rho_L}, \qquad \textit{Equation 58}$$

where ρ_L is the lenticule density expressed in units of lpi.

Quite often N_r is used as the length of the lenticular sequence. With this choice, Equation 58 may give N_r values compatible with matched grids for 600dpi printers (Table 6). For 720dpi printers it is convenient to use the formula

$$N_r = \frac{720}{\rho_L}. \qquad \textit{Equation 59}$$

This gives a slightly larger value of N_r, but the difference between the two estimates is not significant, and Equation 59 may yield values of N_r compatible with matched grids for 720dpi printers (Table 7).

Judging from Figure 24, it is reasonable to assume that rasterization blur reduces N_r by a factor of approximately 2. Thus, in the presence of such blur, the estimates given by Equation 58 and Equation 59 should be reduced by a factor of 2.

6.7 The imaging boxes

The value of h' for $N = N_r$ is an essential parameter in 3D lenticular imaging, and is denoted by h^*:

$$h^* = \frac{N_r t}{n} \quad \text{Equation 60}$$

As argued above, using sequence lengths higher than N_r results in blurring, as adjacent sequence images blend into each other. With $N = N_r$, any point with a distance $h < h^*$ from the picture plane will be displayed both sharply and continuously. Unlike h', the parameter h^* depends only on the lenticular sheet and the printer properties (in the absence of rasterization blur) and does not depend on the sequence length N.

Table 15 gives the estimates of N_r and h^* for a few standard lenticular sheets using Equation 60. The number of resolvable images was calculated using Equation 58, taking the value of 1.6 for the refraction index n. The total parallax values corresponding to h^* are given in the right-hand column.

Table 15: Depth display characteristics for several common lenticular sheets.

t (mm)	lpi (1/inch)	N_r	h^* (mm)	P (mm)
0.3	100	6	1.1	1.3
1	70	9	5.6	2.9
1.3	60	10	8.1	3.8
2	40	15	18.8	8.9
4	24	25	62.5	25.4
4	20	30	75	36.8

Figure 57 introduces an important concept for the understanding and the design 3D lenticular imaging: the resolved imaging box. This is an imaginary box surrounding the lenticular picture. Its width and height are equal to the width and height of the picture, and its depth is $2h^*$. The picture plane is positioned at the center of this box, as shown in Figure 57. Any object imaged within the resolved imaging box will be displayed both sharply and continuously if the sequence length is N_r (or higher). The total depth which can be displayed within this box is $2h^*$.

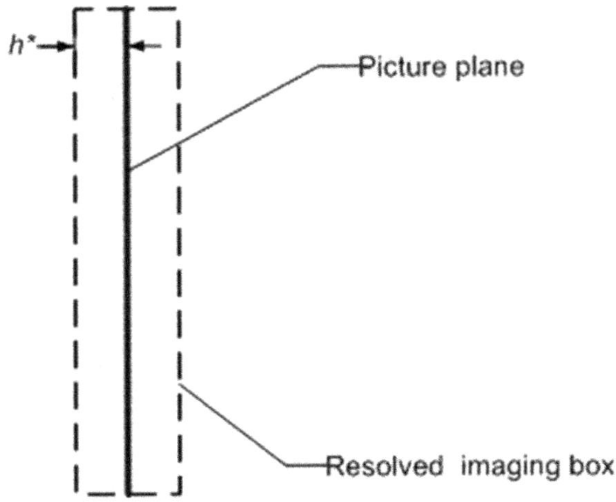

Figure 57: The 3D picture and its resolved imaging box.

With $N = N_r$, objects imaged outside the resolved imaging box may appear as discrete. The magnitude of the discretization increases with the amount of protrusion of the imaged object from the resolved imaging box. However, according to Equation 57, any imaged point with a given distance h from the picture plane can be displayed continuously by increasing N sufficiently. If $h > h^*$, N will need to be increased beyond N_r, introducing blurring. Thus, outside the resolved imaging box any object will appear either blurred or discretized (or both). Inside the resolved imaging box, the blurring will be minimal, or even unnoticeable.

It is convenient to introduce a second imaging box, called the "continuous imaging box". This is a box similar to the resolved imaging box, with the only difference being that its depth is $2h'$ instead of $2h^*$. The depth of the continuous imaging box depends linearly on N, according to Equation 56.

Let us introduce the parameter y,

$$y = \frac{N}{N_r} , \qquad \text{Equation 61}$$

and call it "sequence length factor." It is easily verified that

$$h' = yh^* . \qquad \text{Equation 62}$$

The value of this factor can be chosen at will. However, to fully exploit the imaging resources of the picture N should be greater or equal to N_r. Correspondingly $y \geq 1$ and the depth of the continuous imaging box is greater or equal to the depth of the resolved imaging box. The two imaging boxes are shown schematically in Figure 58. Any object imaged within the continuous imaging box will appear continuous. However, objects within its volume but outside the resolved imaging box will appear blurred.

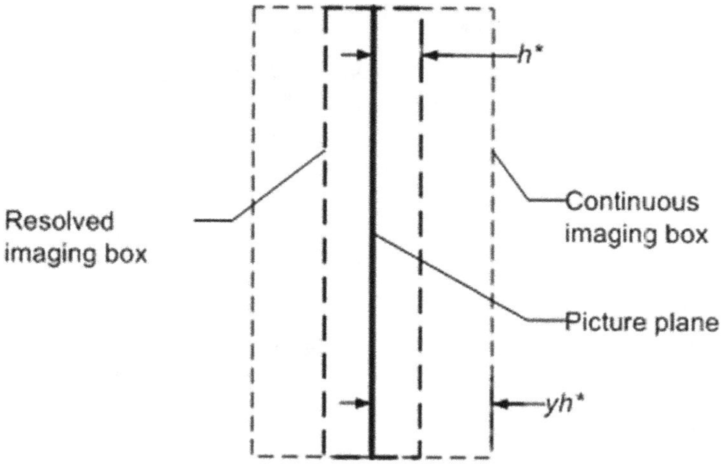

Figure 58: The imaging boxes of a 3D lenticular picture.

The imaging characteristics of the different regions for $N > N_r$ are summarized in Table 16.

Table 16: Imaging characteristics of different regions in the imaged volume.

Position	Characteristics
Inside the resolved imaging box	Sharp and continuous
Inside the continuous imaging box but outside the resolved imaging box	Continuous and blurred
Outside the continuous imaging box	Blurred and discretized

These imaging characteristics have features which are similar to the familiar optical imaging, as experienced by photographers. Photographic lenses have a certain depth of field, which marks a region in the object space in which objects can be imaged sharply (or, "in focus"). The lenticular analog of the depth of field in photographic lenses is the resolved imaging box. Although the mechanisms which give rise to focused, or sharp zones in the two technologies are entirely different, the visual outcome is very similar. There is no analog of the lenticular discretization phenomenon in photography.

Both blurring and discretization are severe visual defects; outside the resolved imaging box one defect can be traded for the other, but it is impossible to eliminate both. In practice, one seeks a certain compromise between the two. The optimal choice for such compromise depends on the picture graphical content and the preferences of the artist. It should also be borne in mind that increasing N increases the processing time and the storage requirements required to compute the interlaced image, which creates a motivation to keep N as low as possible.

6.8 The onset of discretization

Discretization builds up gradually for imaged objects outside the continuous imaging box as their distance from its boundary increases. Near this boundary, it may be hardly noticeable. Therefore, one may increase the displayed depth without causing disturbing discretization defects by protruding outside the continuous imaging box to a certain extent. The magnitude of this protrusion will be measured by the factor x:

$$x = \frac{P_S}{p} \quad \text{Equation 63}$$

This factor will be called "parallax factor." Using Equation 55 and Equation 56, the imaged object point distance from the picture plane can be expressed as follows:

$$h = xh' \quad \text{Equation 64}$$

Thus, x measures the position of the imaged object point in units of h'.

Table 17 gives qualitative estimates of typical picture discretization for different values of the parallax factor. This table should be regarded only as a qualitative guide; in critical applications, it may be necessary to print test pictures with different values of x and visually choose the most suitable one.

Table 17: Discretization impact for various values of the parallax factor.

Parallax factor value	**Discretization impact**
Less than 1	None
Between 1 and 2	None to acceptable
Between 2 and 4	Acceptable to disturbing
Larger than 4	Disturbing, usually unacceptable

Figure 59 illustrates how discretization develops as the value of the parallax factor increases. It shows four photographs of lenticular pictures made with parallax factor values of 1, 2, 4, and 6, all with the same value of N. The objects in this picture protruding most are encircled by a circle in the central bottom part, and the parallax factor corresponds to these objects.

For $x = 1$ the protruding objects are displayed sharply and distinctively. The display quality is still quite good for $x = 2$. For $x = 4$ the discretization is already significant, and for $x = 6$ it becomes unacceptable.

The pictures shown in Figure 59 have lenticule density of 60lpi and sequence length of 8. This is slightly less than the estimated number 10 for resolvable images as given by Equation 58.

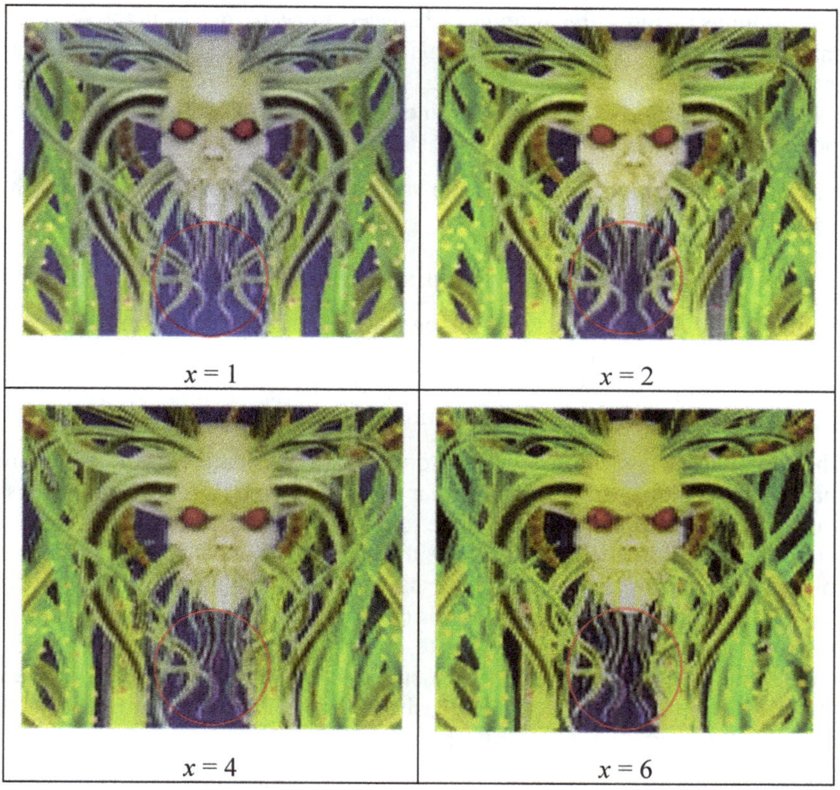

Figure 59: The onset of discretization[3].

The reader may wonder how discretization can be illustrated in a photograph, since it is apparent only when the observer moves his eye so that he can see the transition between adjacent images in the sequence, and this can be seen only while observing the real lenticular picture. The reason that discretization can be seen in a photograph is that when displaying a given image, the lenticular picture will always display traces of the two adjacent images. These traces, sometimes called "ghosts," become more noticeable as the sequence length approaches N_r.

Another way to describe this phenomenon is to say that at each viewing point a lenticular picture displays a weighted superposition of three adjacent images from the sequence. The biggest weight usually belongs to the central image, and the smaller weight to the other two. If the sequence length exceeds N_r, the superposition may consist of even more images. This phenomenon was already illustrated in the measurement of the number of resolvable images (Figure 56). It is manifested in the fact that in all bands one can see more than one bar with different gray levels.

[3] Courtesy of Pop3DArt

In continuous pictures, the differences between adjacent images in the sequence are too small to be noticeable, and therefore this superposition does not affect the image appearance. This is illustrated in the image labeled $x = 1$ in Figure 59. The ghosts become more noticeable as the value of x increases, and for $x = 6$ they significantly degrade the picture display in the region inside the circle at the bottom.

The displayed depth increases with the value of x, but this cannot be seen in the photographs shown in Figure 59.

6.9 Adjusting an object for display in a 3D lenticular picture

The depth design process starts with a given lenticular sheet and a given 3D object. The lenticular sheet is shown in Figure 60 together with its resolved imaging box. The object in this example consists of three sub-objects: the central one is in front, and the other two are behind it. The viewing direction is from the right.

The lenticular picture offers viewing from a relatively small range of angles around the main viewing direction; the total angular viewing interval is limited to the lens viewing angle. Therefore, it is not possible to see the whole surface of an imaged object. In the example shown in Figure 60, the visible part of the surfaces is denoted by bold curves, and this annotation will be used in the following figures as well. The exact extension of the visible parts depends on the viewing angle. The indication of the visible surfaces in Figure 60 is approximate, and its only purpose is an illustration. Also, the front sub-object will partially obscure the two sub-objects located behind it. This effect is ignored in Figure 60. Obviously, in the 3D imaging design, one needs to be concerned only with the visible surfaces.

The visible surfaces of the object in Figure 60 are shown surrounded by a box which is called the “object box." This is the smallest box which contains all visible surfaces of the object. Distinguishing between the real and the imaged objects is essential. Figure 60 and Figure 61 show the real object box, which is completely separated from the resolved imaging box. It is shown next to the picture, but it can be located anywhere.

It is evident that to display the object, it needs to be scaled to fit the picture aperture. Such a scaled object is shown in Figure 61. It may be that the object is much smaller than the sheet aperture, and in such a case one may choose to upscale the object. The depth of the scaled object is denoted by D.

The picture may be designed so that the picture plane is at the center of the imaged object box, as shown in Figure 62 (a). In this arrangement, the visual defects will have the same magnitude in the front and the rear. However, it is also possible to display the object box asymmetrically, as shown in Figure 62 (b). In this example, the picture plane is shown closer to the front facet of the imaged object box, so that the visual defects will be more conspicuous in the rear compared to the front. Such geometry may be useful if for some reason the receding objects are more tolerant to visual defects than the front objects, or if they are less important for the composition. For instance, in a portrait picture, the face of the subject will be more important than the background.

To simplify the discussion, it will be assumed from now on that the imaged object box is centered with respect to the picture plane (Figure 62a).

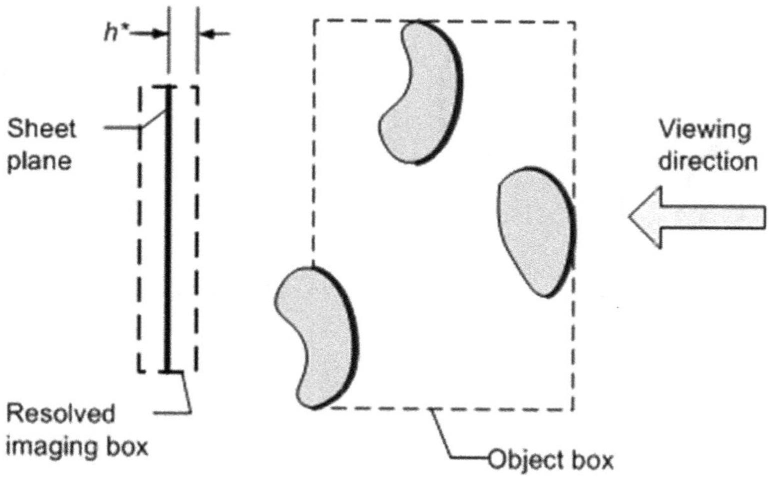

Figure 60: The initial state for 3D imaging design.

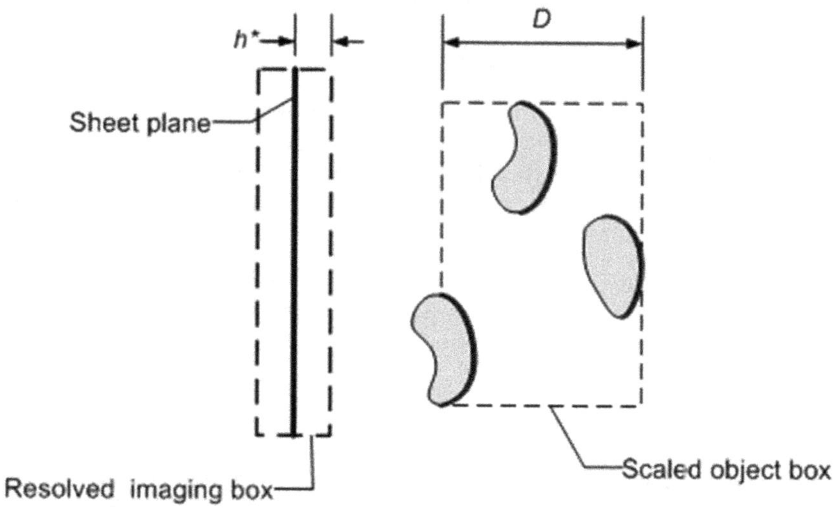

Figure 61: Scaling the object to fit the sheet aperture.

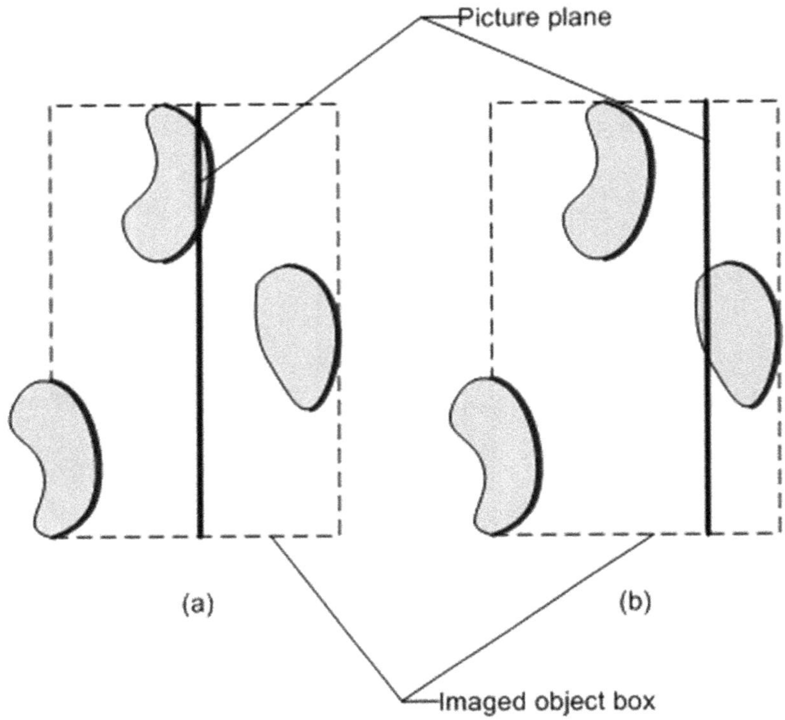

Figure 62: Display of the imaged object box relative to the picture plane. (a) symmetric, (b) asymmetric.

6.10 The depth enhancement factor

In the example shown in Figure 61, the object depth D is larger than the resolved imaging box depth $2h^*$. Displaying it with its original depth will introduce visual defects. To avoid this, the displayed depth may be reduced to fit inside the resolved imaging box. The object will appear squeezed, as shown in Figure 63, but without visual defects.

Quite often there is a high motivation to display more depth, even at the cost of introducing visual defects. The extent by which the depth display is increased beyond the resolved imaging box will be measured by the parameter z:

$$z = \frac{D}{h^*} \quad \text{Equation 65}$$

This parameter is called the "depth enhancement factor." One may display the object in its correct proportions, as shown in Figure 61. One may even increase the displayed depth of the object by distorting its proportions. In the correct object proportions shown in Figure 61, the depth enhancement factor is approximately 2.

Using Equation 62 and Equation 64, z may be expressed as follows:

$$z = xy\ ,$$ *Equation 66*

where x corresponds to the most protruding (or most receding) visible object point.

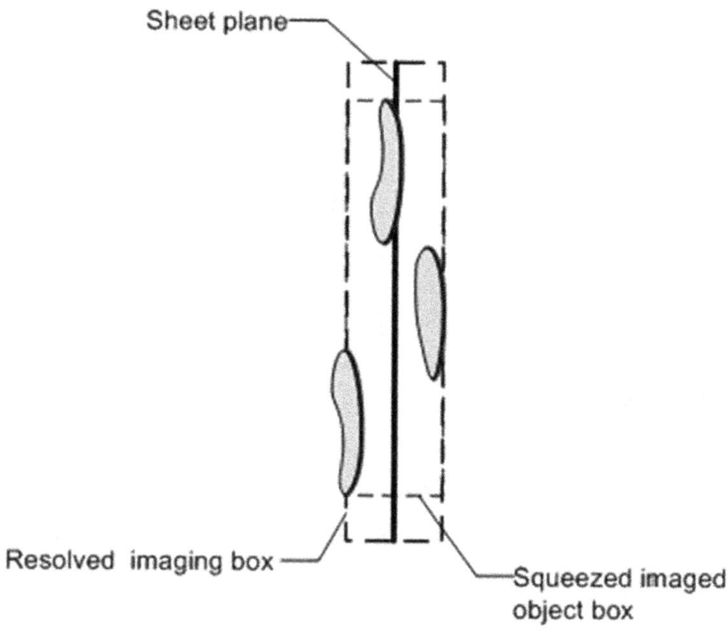

Figure 63: Squeezing the object to fit the resolved imaging box.

6.11 3D imaging design

The design process usually starts with a decision on the value of z. If a value greater than one is chosen, the blend of the visual defects which will be introduced can be chosen by controlling the value of x or y. By assigning a certain value to one of these parameters, the value of other one is determined from Equation 66. Alternatively, one may choose the values of x and y, and then z is determined by Equation 66. Let us consider, for example, the possible designs with $z = 2$.

Figure 64 shows an extreme design choice with $y = 1$, in which case the resolved and the continuous boxes coincide. The x value for the most protruding (and most receding) point will be 2. According to Table 17, this value of x is usually permissible, so that the design shown in Figure 64 is valid.

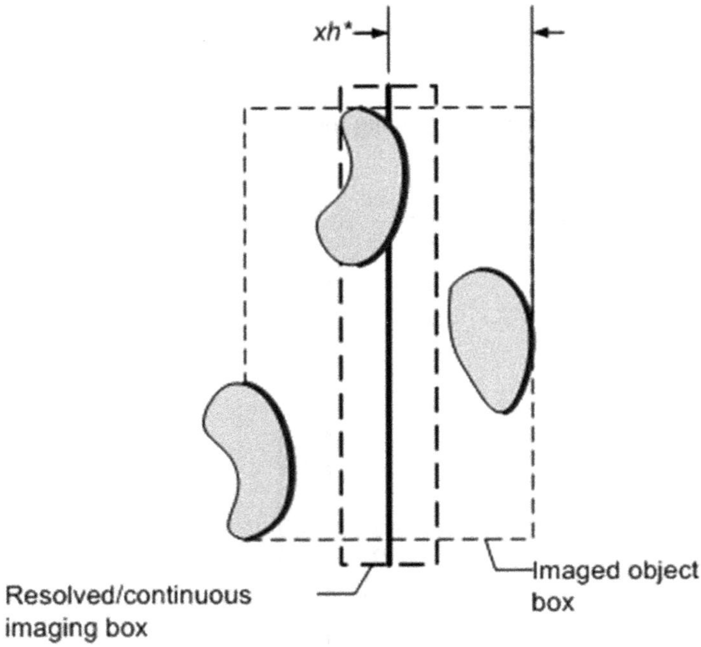

Figure 64: 3D imaging without blurring (y = 1).

The other extreme design case corresponds to choosing $x = 1$ and is shown in Figure 65. In this design, the whole object box is contained within the continuous imaging box. Therefore, there is no discretization. However, the regions outside the resolved imaging box may appear blurred to a certain degree.

The general 3D imaging design with $x > 1$ and $y > 1$ is shown in Figure 66. The object parts outside the resolved imaging box will be partly blurred. The object parts beyond the continuous imaging box will be both blurred and discretized to a certain degree.

Table 18 summarizes the sequence characteristics for the three designs discussed here and a given depth enhancement factor z. The incremental parallax P_S of a given object point depends on its distance h from the picture plane. At the picture plane, P_S vanishes. For each design, two values of P_S are listed in Table 18: at the edge of the resolved imaging box ($h = h^*$) and its maximum. As argued, continuity requires P_S to be less or equal to p. Therefore, only the first design will display a fully continuous picture.

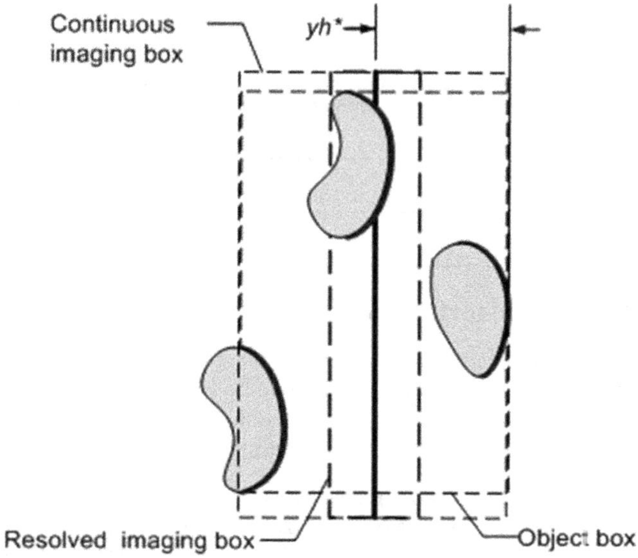

Figure 65: 3D imaging without discretization (x = 1).

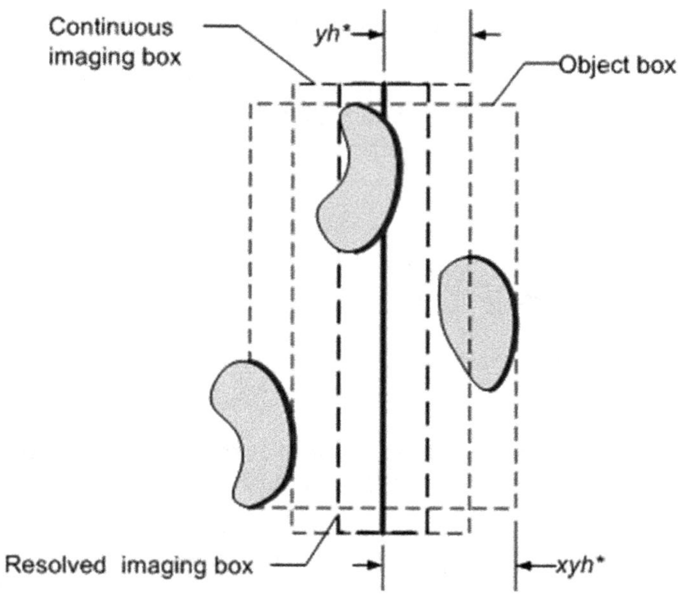

Figure 66: The general 3D imaging design.

Table 18: Values of N and P_S for a given z.

Case	x (max)	y	N	P_S @ h^*	P_S (max)
Fully continuous	1	z	zN_r	p/z	p
Fully sharp	z	1	N_r	p	zp
General	z/a	a	aN_r	p/a	zp/a

6.12 The effect of rasterization blur

Until now it was assumed that the number of resolvable images is constant across the picture area. This is characteristic of a perfectly matched or a greatly unmatched pixel grid. A lack of matching introduces rasterization blur, which affects the number of resolvable images in a spatially non-uniform manner.

As noted in paragraph 6.6, the number of resolvable images is reduced by a factor of 2 (approximately) in the presence of rasterization blur. Since this blur is a periodic function of the horizontal picture coordinate (denoted by q), the number of resolvable images is periodic too and oscillates along the horizontal direction between N_r and $N_r/2$.

When the pixel grid is greatly unmatched, the oscillation period is small relative to the picture width, and there are many oscillations within the picture. On the other hand, if the grid is nearly matched, the number of oscillations is small, or even there are no oscillations at all. Figure 67 shows the dependence of the number of resolvable images on q for the case of a single oscillation period schematically. The maximum of the curve appears in the picture center, which is characteristic of a centered grid (3.4). The position of the maximum can be shifted by shifting the pixel grid from the central position.

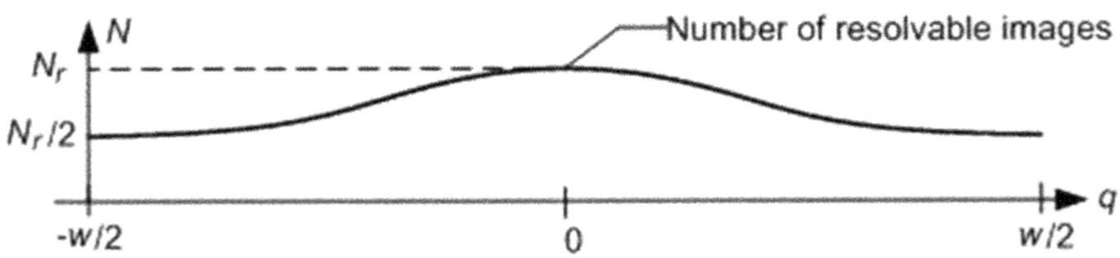

Figure 67: The variation of the number of resolvable images along the horizontal coordinate for a nearly matched pixel grid

When the pixel grid is nearly matched, the resolved imaging box assumes a non-rectangular shape and should be called the "volume" rather than "box." Such volume is shown schematically from the top in Figure 68 for the single-period and centered grid case of Figure 67. In such a volume, there is an enhanced capability of depth imaging at the picture center, compared to its edges.

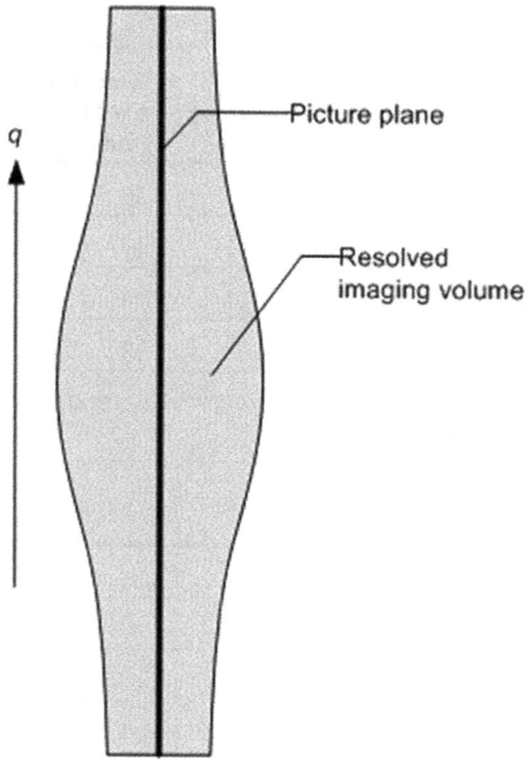

Figure 68: The resolved imaging volume of a picture with a nearly matched pixel grid (top view).

6.13 Depth display design practice

It is hard to find published guidelines for choosing x and y. Knigin (Knigin, 2016) appears to suggest setting the largest incremental parallax to p. This causes the imaged object box to be contained in the continuous imaging box, so there will be no discretization. In the present terms, this corresponds to choosing $x = 1$.

Dean (Dean, 2015) suggests a total parallax of 76mm and 38mm for receding imaged objects in 20lpi and 40lpi lenses respectively. For protruding imaged objects, he suggests half of these values. In both cases, Dean recommends using $N = 36$. Dean's recommendations are summarized in Table 19. In the following table, the calculated values of the depth parameters values are shown, assuming a thickness of 2mm for the 40lpi sheet and 4mm for the 20lpi sheet.

Table 19: Dean's picture design recommendations for 40 lpi and 20lpi sheets.

Case no.	Position	lpi (inch^{-1})	p (mm)	Recommended total parallax (mm)	P_s (mm)
1	Protruding	20	1.27	38	1.05
2	Protruding	40	0.635	19	0.53
3	Receding	20	1.27	76	2.1
4	Receding	40	0.635	38	1.05

Table 20: The values of the depth parameters x, y, and z for Dean's recommendations.

Case no.	Position	lpi (inch^{-1})	x	y	z
1	Protruding	20	0.83	1.2	1
2	Protruding	40	0.83	2.4	2
3	Receding	20	1.65	1.2	2
4	Receding	40	1.65	2.4	4

6.14 Validity of the small parallax approximation

Finally, let us verify that our estimates are compatible with the small parallax approximation (which was used to derive them). The largest value of h^* is in the last row, corresponding to 143mm. Taking a depth enhancement factor of 2, pictures made with this lenticular sheet will be designed with a maximal object distance of about 286mm. Such pictures are usually rather big and are viewed from distances of 2000mm or more, which are indeed much greater than 286mm. Smaller pictures may be viewed from shorter distances, but they are usually made with much finer lenticular sheets, whose values of h^* are smaller.

In extreme cases where z is chosen to be significantly greater than 2 or the picture is viewed from an unusually close distance, some deviations in the displayed depth may occur, and for correct estimates the rigorous formulas (Equation 52 and Equation 53) must be used.

Chapter 7
Electronic lenticular displays

"Engineers are sometimes the least practical of men, they may be attracted by difficulties"

-David McCullough, in "The Path Between the Seas: The Creation of the Panama Canal, 1870-1914" (attributed to an anonymous railway contractor)

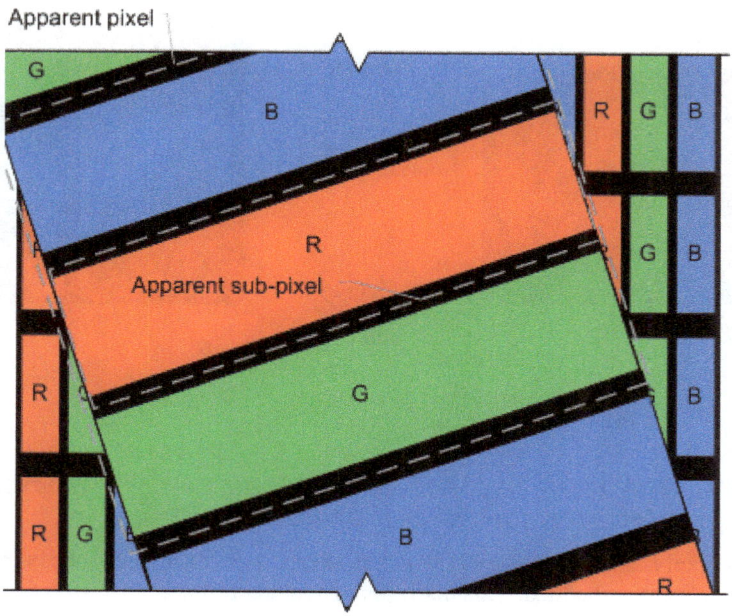

7.1 Introduction

The backplane picture in electronic lenticular displays is created by a flat-panel display, like an LCD screen. The use of such screen allows changing the backplane picture, and even the playback of a video, thus creating a live and glass-free 3D display. It was anticipated that such display would be attractive as a TV set, and therefore much resources were invested in developing it.

The pixels of electronic displays are made from three monochromatic sub-pixels, each colored in one of the three fundamental colors: red, blue, and green (Byoungho, Soon-gi, Keehoon, & Jisoo, 2016) (Leuder, 2012). The size of the sub-pixels is made to be smaller than the eye resolution, so visually their colors blend and full color can be displayed for each pixel. In all displays, the pixels form a rectangular array, as in print. A common sub-pixel structure in LCD displays is shown in Figure 69. Other sub-pixel structures exist, but in this chapter, the structure shown in Figure 69 will be assumed.

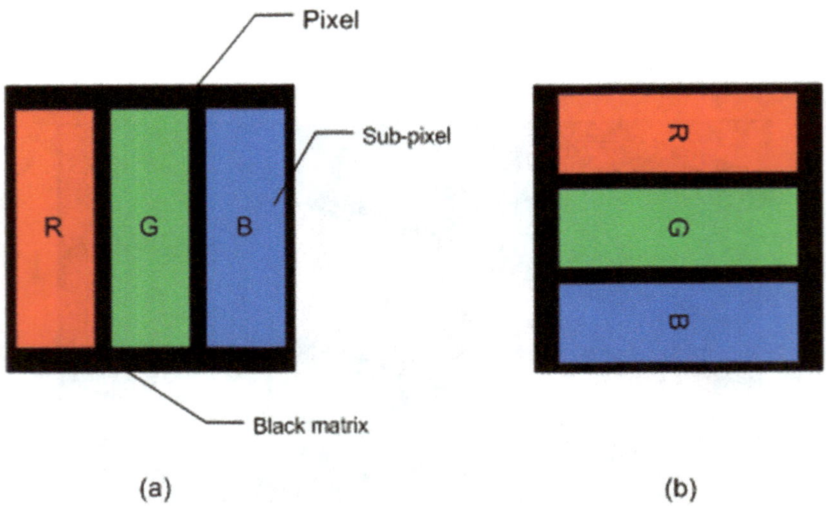

Figure 69: Pixel structure of an LCD display: (a) display in landscape orientation, (b) display in portrait orientation

The LCD pixel has another element which is absent in print: a black matrix in which the sub-pixels are embedded. This matrix is necessary for the optical isolation of the sub-pixels, but in lenticular 3D displays, it creates a visual defect known as "Moiré fringes" (van Berkel & Clarke, Characterisation and Optimisation of 3D-LCD Module Design, 1997).

Unlike print, in which the pixel density can be specified at will, the pixel density in electronic displays is fixed, and known to an excellent precision.

7.2 Slanted displays

Figure 70 shows a single pixel column of a display in landscape orientation. The broken lines denote optical gridlines in two positions for a lenticule with an axis parallel to the pixel gridlines.

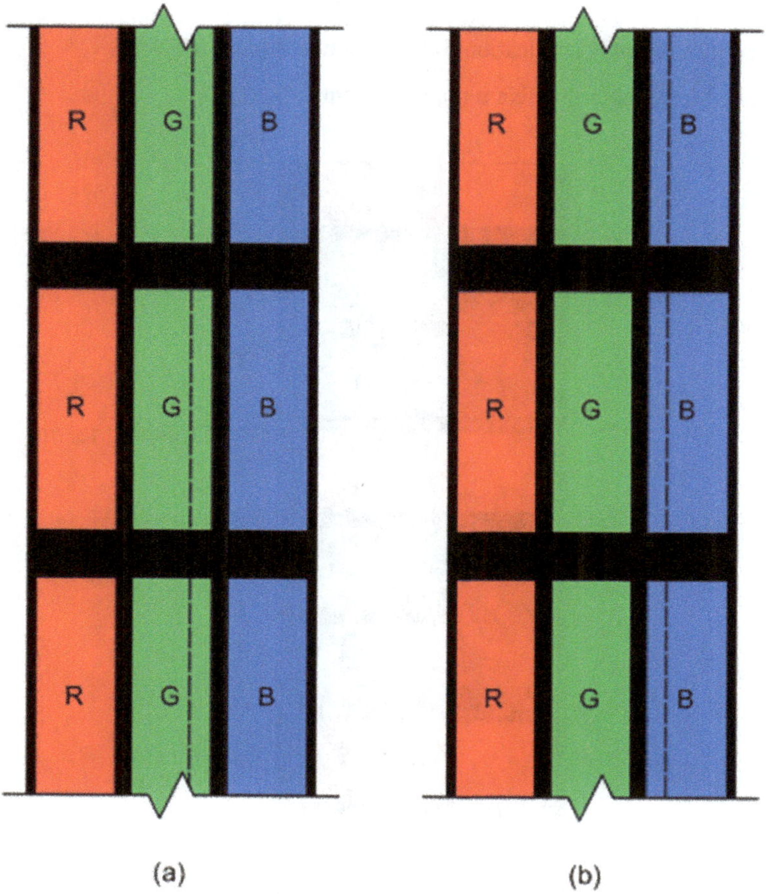

Figure 70: Electronic lenticular display in landscape orientation with a vertically aligned lenticular sheet. (a) only green color is displayed, (b) only blue color is displayed.

In case (a), the optical gridline overlaps green sub-pixels only, and therefore the corresponding lenticule will show only green color. In case (b), the optical gridline overlaps blue sub-pixels only, causing only blue color to appear in the corresponding lenticule. This is unacceptable; it must be possible to display full color in each lenticule.

One way to solve this problem is to use the display in portrait orientation. Figure 71 shows an electronic display in portrait orientation, with an optical gridline superimposed. In this case, the optical gridline overlaps sub-pixels in all three primary colors, and therefore can be painted in any color. However, this solution has two disadvantages:

1. The display is constrained to be in portrait orientation, which is less attractive for most display applications, and most notably, for TV sets,
2. The Moiré pattern defect remains present.

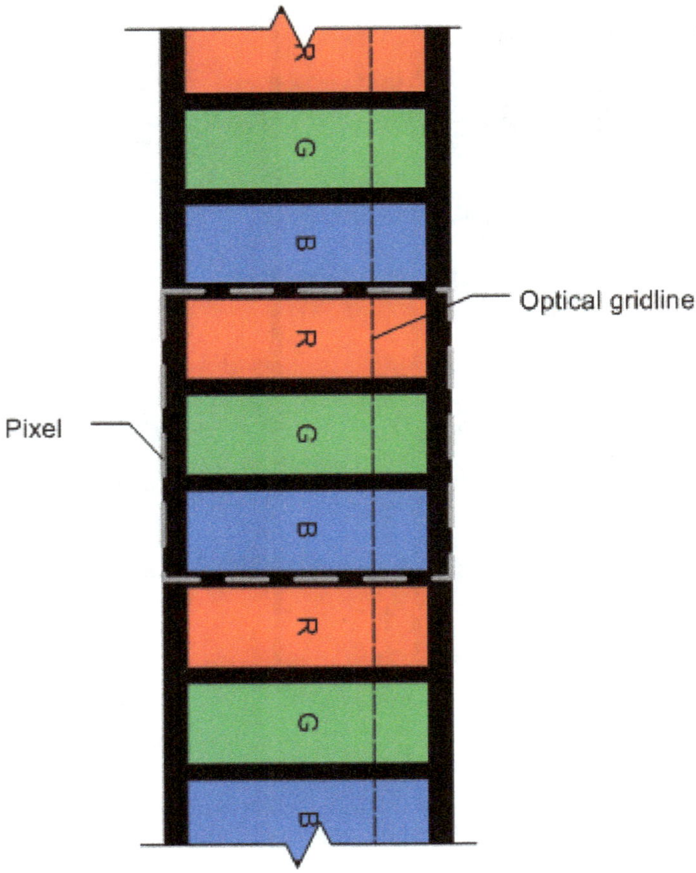

Figure 71: Optical gridline in portrait oriented display

Both problems are solved by introducing an angle between the lenticular grid and the display pixel grid, known as "slanting" (van Berkel & Clarke, Characterisation and Optimisation of 3D-LCD Module Design, 1997). In such displays, the optical grid will also be slanted with respect to the pixel grid, as shown in Figure 72. In this figure, the tangent of the slanting angle is 1/3 (the angle itself is approximately 18.43°, measured from the vertical), but other angles may be used. It is seen that a slanted optical gridline will overlap sub-pixels of all three colors. Since colors of adjacent sub-pixels blend visually, slanting allows full-color display for each given lenticule. It turns out that slanting also reduces the Moiré effect, and for certain angles eliminates it completely (van Berkel & Clarke, Characterisation and Optimisation of 3D-LCD Module Design, 1997).

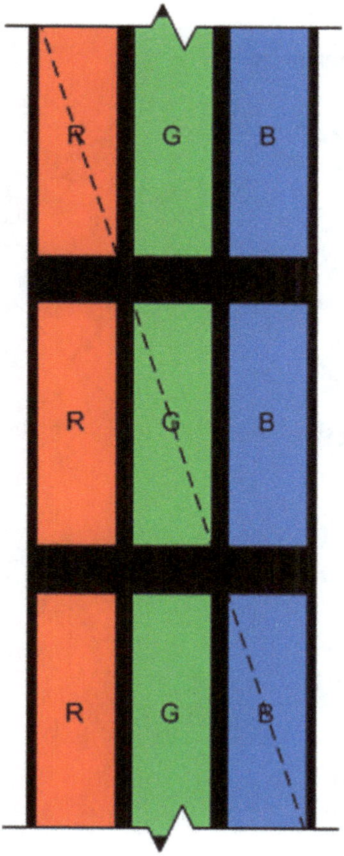

Figure 72: Slanted optical gridline.

In Chapter 3 through Chapter 6 it was implicitly assumed that the lenticules are parallel to the pixel grid of the backplane picture. This is the case in printed lenticular pictures. Such pictures will be called in this chapter "erect" as opposed to "slanted." Erect pictures are a special case of slanted pictures, with the slant angle equal to zero.

7.3 Optical effects of a slanted lenticule

7.3.1 The effect of a slanted lenticule on a line

In Chapter 2 it was shown that the graphical content that is contained in the visible zone of a lenticule is expanded optically to fill its whole aperture. The expansion is in the direction which is perpendicular to the lenticule axis. This effect is illustrated in Figure 73, for a horizontal lune covered by a slanted lenticule.

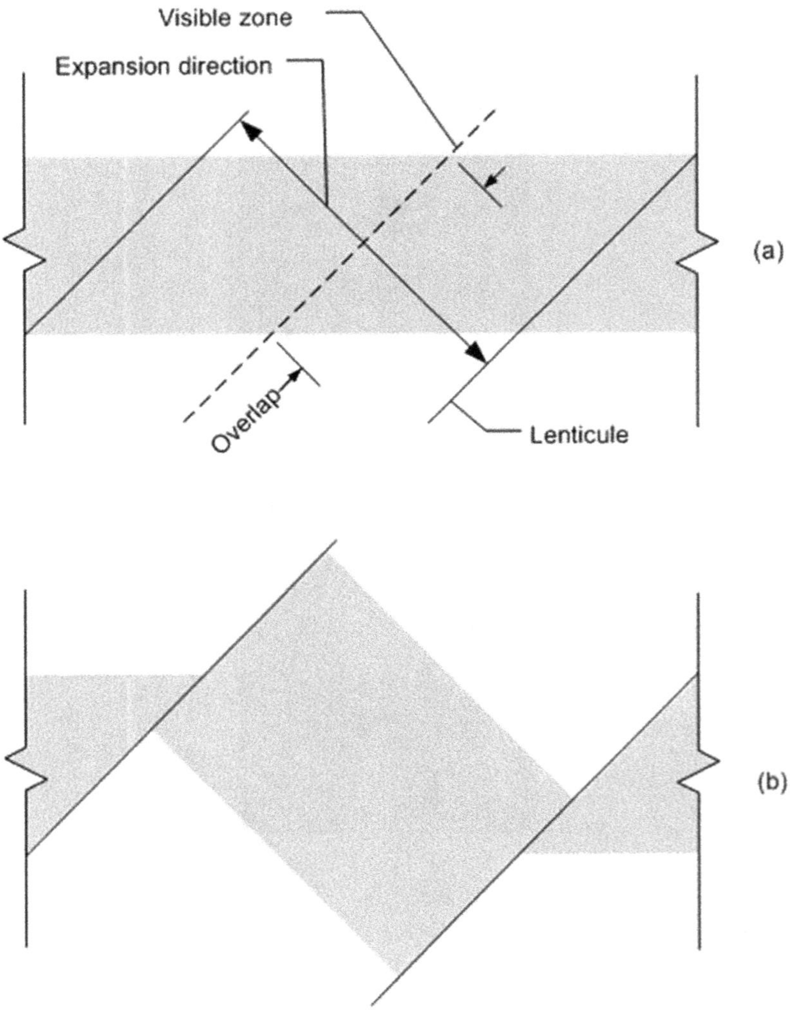

Figure 73: The visual effect of a slanted lenticule on a horizontal line.

Part (a) of Figure 73 shows a slanted lenticule superimposed on a horizontal gray line on a white background. The line is shown without the optical effect of the lenticule, and the slanted solid lines denote the boundaries of the lenticule. Its visible zone is much narrower than its width and is represented by a broken line. Its section overlapping the gray line is expanded optically to the lenticule boundaries, forming a rectangle, as illustrated in part (b) of Figure 73. This effect is further illustrated by a real photograph in Figure 74.

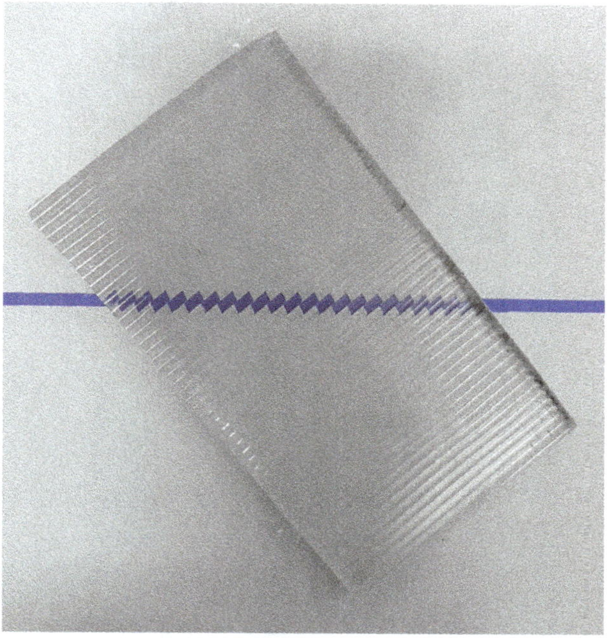

Figure 74: The visual effect of a slanted lenticular sheet on a horizontal line: real photograph.

7.3.2 *Apparent pixels in slanted displays*

The concept of apparent pixels was introduced in paragraph 2.8.3 to describe the optical effect of a lenticular sheet on a picture, when the lenticules are parallel to the pixel grid. Apparent pixels are created in slanted displays too. However, their geometry is significantly different.

To illustrate the effect of a slanted lenticule on a pixel grid, let us consider a special raster image which is made of three kinds of pixels, painted white, gray, and black, and arranged as shown in Figure 75. The superimposed slanted lenticule with its visible zone is also shown, without the optical effect. The slant angle is 45°, and the visible zone passes through the pixels vertices. This position is called "diagonal."

Figure 76 shows how the pixel grid looks like with the optical effect of the lenticule. The apparent pixels become rectangles also slanted by 45°. The apparent pixels width is the same as the lenticule width, and their height is equal to the diagonal size of the original pixels. In the diagonal position, each apparent pixel displays graphical content from a single original pixel, like in the case of the non-slanted lenticule.

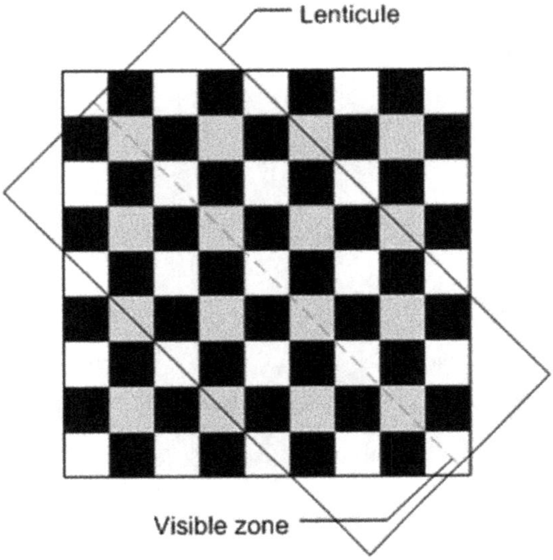

Figure 75: Slanted lenticule superimposed on a pixel grid with a diagonal visible zone.

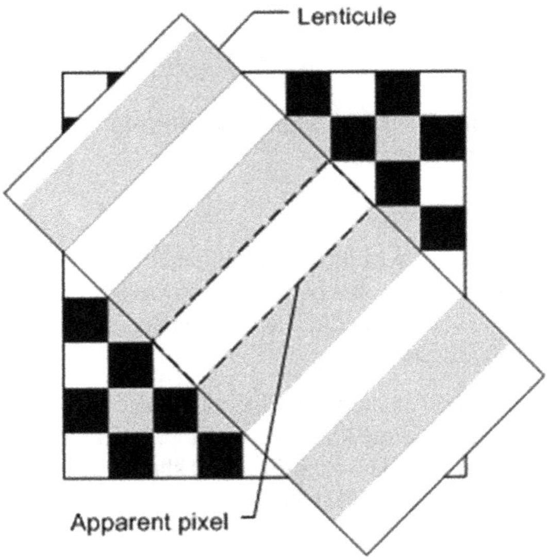

Figure 76: The apparent pixels in a slanted lenticule with a diagonal visible zone.

Figure 77 and Figure 78 show the general case, in which the visible zone does not pass through the original pixels vertices (also the slant angle remains at 45°). The optical effect is shown in Figure 78, and it is seen that now each apparent pixel is split into two rectangles, each having a different gray level. This results in blurring. The heights of the rectangles are, in general, not equal, and depend on the exact position of the visible zone.

7.4 Pixel structure in slanted electronic displays

The method explained in the previous paragraph can be used to derive the geometry of the apparent pixels in the slanted electronic display. For clarity, the illustrations will be for a single lenticule.

Let us consider the case in which the visible zone is in the diagonal orientation. For a display with sub-pixels, it is convenient to regard a viewing zone which crosses the sub-pixel vertices as diagonal. Such viewing zone is shown in Figure 72. Figure 79 shows a lenticule with a diagonal viewing zone superimposed on a pixel grid of an electronic display without the optical effect. The optical effect of the lenticule is shown in Figure 80 and is manifested by the appearance of the apparent sub-pixels. It is seen that the apparent sub-pixels are magnified and slanted with respect to the display sub-pixels, and their aspect ratio is modified. A sequential group of R, G, and B sub-pixels forms the apparent pixel. Note that the black matrix is present also in the apparent pixels.

The geometry of the apparent sub-pixels depends on the position of the visible zone. In a general position, each sub-pixel is split into two sub-pixels, but the shape of the apparent pixel remains intact, and it will contain equal areas of the three primary colors. Therefore, all apparent pixels can display full color.

It is to be noted that the aspect ratio of the apparent pixels in slanted displays is much smaller than in erect pictures (like the printed lenticular pictures). In fact, in the example shown in Figure 79 and Figure 80, the apparent pixel is almost a square. This distributes the loss of resolution between the horizontal and the vertical directions and therefore increases the efficiency of the display pixels utilization. The increase in efficiency is very significant in LCD displays, in which the available pixels number is the most significant limitation for the 3D display application.

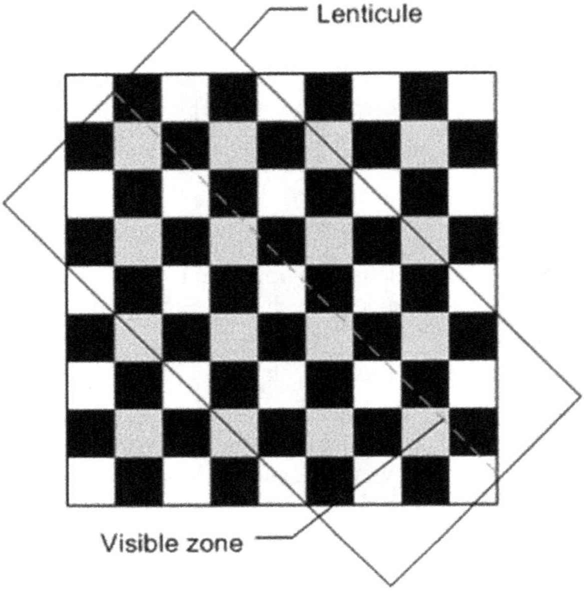

Figure 77: *Slanted lenticule superimposed on a pixel grid with a visible zone in an arbitrary position.*

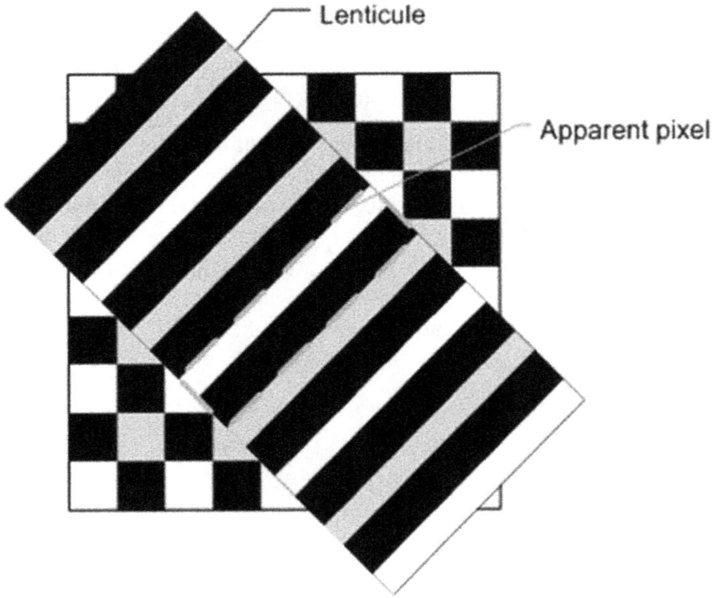

Figure 78: *The apparent pixels in a slanted lenticule with a visible zone in a general position.*

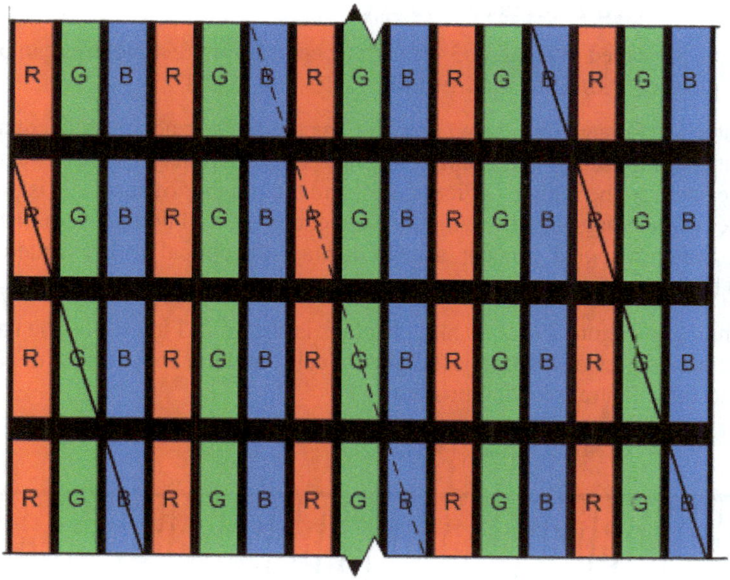

Figure 79: A lenticule superimposed on an LCD panel without showing the apparent pixels: diagonal position.

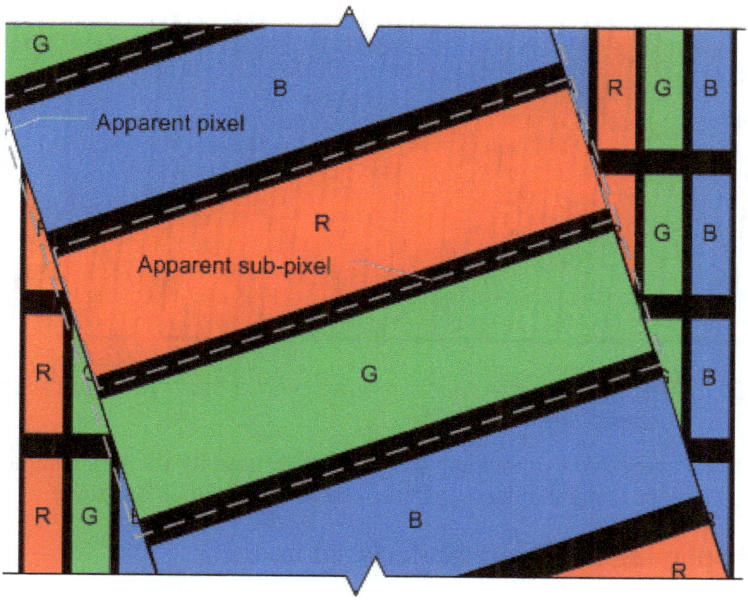

Figure 80: The apparent pixel corresponding to the lenticule of Figure 79.

7.5 The slanted interlaced image

The analysis presented in Chapter 3 through Chapter 6 for erect pictures can be applied to slanted pictures with some modifications.

The interlaced image for slanted displays has the same structure as in erect pictures: it is composed of cells, and each cell is formed from sub-cells. The number of cells is the same as the number of lenticules, and the number of sub-cells is the same as the number of the images in the lenticular sequence. The cell boundaries must be parallel to the optical grid. Since in slanted displays the optical grid is slanted too, so must also be the cells of the interlaced image.

The analog of Figure 29 for slanted displays is shown in Figure 81. The solid lines denote the boundaries of the interlaced picture cells, and the broken lines denote the optical grid for a central viewing position.

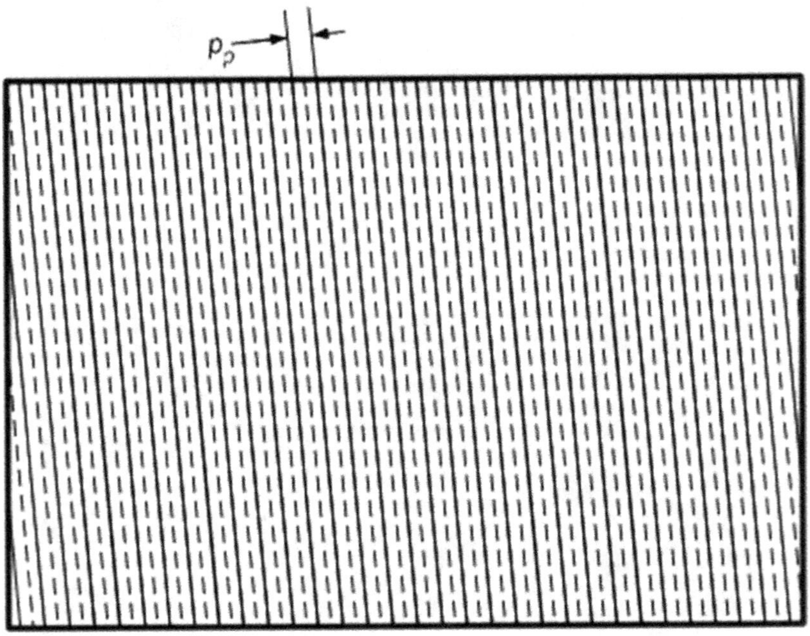

Figure 81: The slanted interlaced picture with a centered optical grid (broken lines).

The algorithm presented in paragraph 4.2 can also be used for the slanted image, with the only difference being that its output must be rotated by the slant angle.

7.6 Displaying the slanted interlaced image

The methodology for printing the backplane picture consisted of three stages: creating the interlaced image, rasterizing, and then printing it. In electronic displays, computational speed is of primary importance, to enable real-time video operation.

Therefore, developers of slanted displays skipped the intermediate stage and developed algorithms which convert the lenticular sequence directly to the backplane picture (van Berkel, Image Preparation for 3D-LCD, 1999). Direct algorithms are more efficient computationally, and therefore are faster. However, it is possible to apply the methodology of erect pictures to slanted displays too, as outlined below.

The cell width of the interlaced picture, denoted in Figure 81 by p_P, must satisfy the same condition as in printed pictures: it must be equal to the optical grid pitch. To match these quantities, it is necessary to measure the pixel size of the optical grid pitch, or, in other words, to perform calibration.

Calibration of slanted displays can be done with the same methods as the calibration of printed pictures. The same calibration patterns which were explained in Chapter 5 can be used, except that they need to be rotated by the slant angle.

Once the pixel size of the cell is determined, it is possible to rasterize the slanted interlaced image so that it can be displayed on the LCD screen. However, since the interlaced image cells are not parallel to the pixel grid, it is not possible to match these grids, and therefore, rasterization blur is unavoidable. This blur reduces the number of resolvable images.

The development of a rasterization algorithm for slanted interlaced images is a subject for future research. It may be advisable to consider the following points for the development of such algorithm:

1. The image should be decomposed into its three color components, and each color component should be rasterized independently with respect to its pixel sub-grid,
2. The apparent sub-pixels geometry should be considered in the sampling algorithm.

7.7 Depth imaging in slanted displays

The optical grid is translated horizontally according to the movement of the observer. Since this grid is periodic, there are certain horizontal translations which leave it invariant. Let us denote by p_S the smallest of these translations. This translation is the slanted interlaced image cell width as shown in Figure 82.

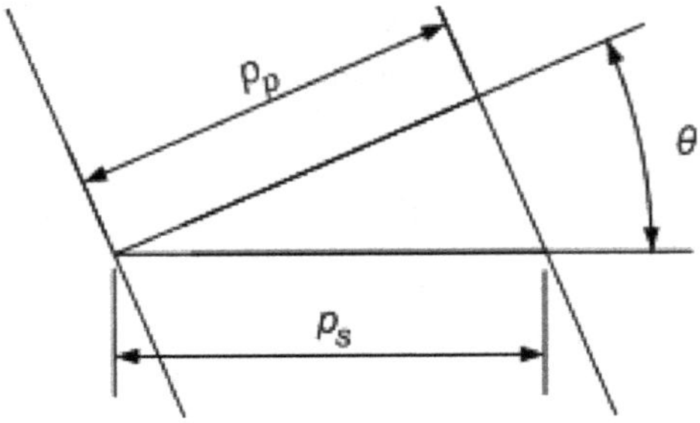

Figure 82: The slanted cell width of a slanted interlaced image

The relation between the interlaced image cell width p_p and the slanted cell width p_s is derived from Figure 82:

$$p_s = \frac{p_p}{\cos\theta}. \qquad \text{Equation 67}$$

Correspondingly, the main viewing interval length (Figure 48) in slanted display is

$$S_s = \frac{S}{\cos\theta}.$$

It is left to the reader to confirm that the quantity h (Equation 55) is transformed to

$$h_s = h\cos\theta = \frac{Nt}{n}\frac{P_s}{p}\cos\theta. \qquad \text{Equation 68}$$

The size of the horizontal resolution element d_s of a slanted display is the horizontal projection of the apparent pixel diagonal, shown by a broken line in Figure 83. The size of d_s can be expressed in terms of p, θ, and the display pixel size. This expression turns out to be quite messy; for the sake of qualitative estimations, it is possible to use the approximation

$$d_s \approx p\cos\theta, \qquad \text{Equation 69}$$

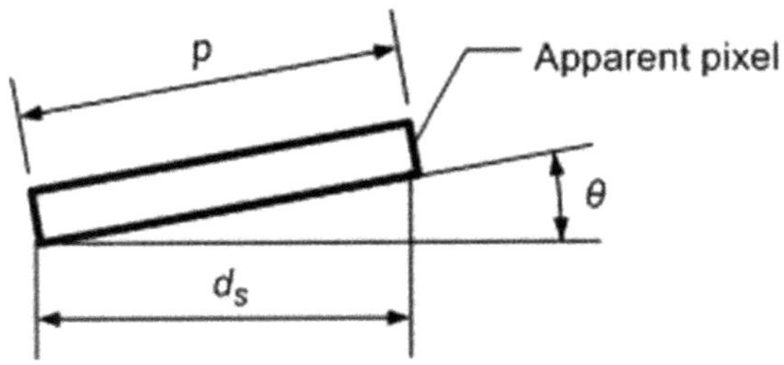

Figure 83: Horizontal resolution element of a slanted display (small slanting angle)

which is valid for $\theta << 1$. A continuous 3D display will be achieved if the incremental parallax P_s is smaller than d_S. Substituting d_s for P_s in Equation 68 yields

$$h'_s = \frac{Nt}{n}\cos^2\theta = h'\cos^2\theta$$

where h'_s is the limit for continuous imaging for the slanted display. Correspondingly, the limit h^*_s for continuous and resolved imaging for the slanted display is modified to (Equation 60)

$$h^*_s = h^*\cos^2\theta \qquad \text{Equation 70}$$

7.8 The number of resolvable images in slanted displays

The number of resolvable images N_r reflects the resolution of the display and is the combined outcome of several mechanisms, one of which being the rasterization blur. In the case of erect images, it is possible to avoid the rasterization blur (paragraph 6.6) by using matched pixel grids (paragraph 3.4). The use of such grids is impossible in the case of slanted displays, in which there are no matched pixel grids, and rasterization blur cannot be avoided. Therefore, any measurement of N_r must be done with its presence.

In the case of printed pictures, it was possible to estimate N_r in the absence of the rasterization blur by dividing the lenticular pitch by the smallest printable linewidth. In an electronic display, the analogous estimate is the ratio between the slanted cell width and the display pixel size:

$$N_r(\text{without rasterization blur}) \approx \frac{p_S}{\text{Display pixel size}}. \qquad \textit{Equation 71}$$

However, this estimate is always reduced by rasterization blur. The amount of this reduction depends on the rasterization method and the slant angle. Rigorous estimation of N_r in slanted displays is subject to future research. Meanwhile, it is reasonable to assume that the rasterization blur reduces N_r by a factor of 2:

$$N_r(\text{with rasterization blur}) \approx \frac{p_S}{2\cdot(\text{Display pixel size})}. \qquad \textit{Equation 72}$$

7.9 Characteristics of TV displays

The present three common formats for TV displays are listed Table 21. The 8K format is new, released commercially only this year (2017) in 70” size.

Table 21: TV formats and pixel dimensions

Format name	Pixel dimensions
HD	1920x1080
4K	3849x2169
8K	7680x4320

The continuously resolved depth display capability of electronic lenticular pictures using such TV displays is much smaller compared to printed pictures, due to the reduced pixel density. Three typical estimates are given in Table 22. The estimates were given for a 10lpi lens with 4mm thickness, and Equation 72 was used to estimate the number of resolvable images N_r (with rounding). Equation 70 was used to derive the values in the rightmost column (again with rounding), which give the depth of the resolved imaging box. The slanting angle was taken as 18.43°.

Table 22: Depth display capability of TV displays with a 10lpi lens.

Format	Size (inch)	N_r	$2h^*_s$ (mm)
HD	40	3	13
4K	40	6	27
8K	70	7	31

These estimates of the depth display capability of TV displays are disappointingly small, even for a lenticular sheet as coarse as 10lpi. These estimates correspond to imaging which is both continuous and sharp. If these quality requirements are discarded, the size of displayed depth can be increased at will.

Chapter 8
The viewing space of 3D lenticular pictures

"If it is gratitude you want, get a dog."

- Robert Harris, in "Imperium" (attributed literary to Cicero)

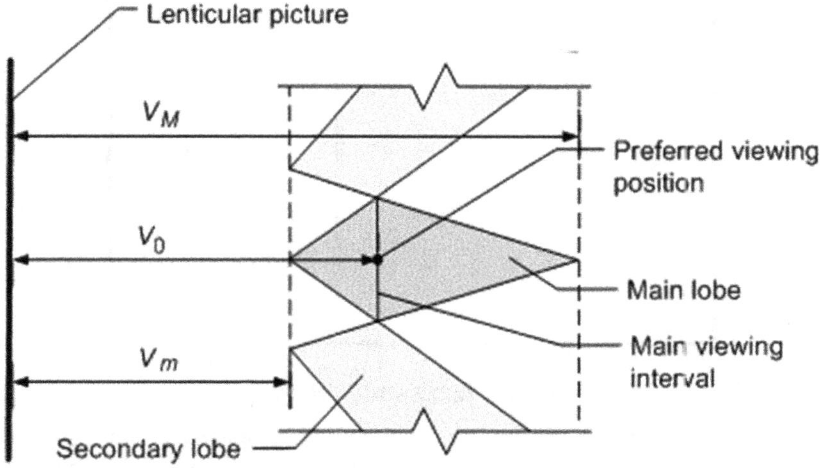

8.1 The break line

The visual properties of a lenticular picture when viewed from the calibration distance were discussed in paragraph 4.4. At this range, the picture displays a single image from any viewing point. However, lenticular pictures are observed most of the time from other distances. At a viewing distance which is different from the calibration distance, there will be in general viewing positions in which the picture will exhibit visual defects. A view showing the picture without visual defects will be referred to as "proper view."

When the viewing distance differs from the calibration distance, the picture may display two or even more successive sequence images in different parts of its area. In a perfectly aligned picture, the regions showing single images are rectangles of the same height as the picture height. The borders between these areas are called "transition lines." A scheme of such a picture displaying two successive sequence images is shown in Figure 84.

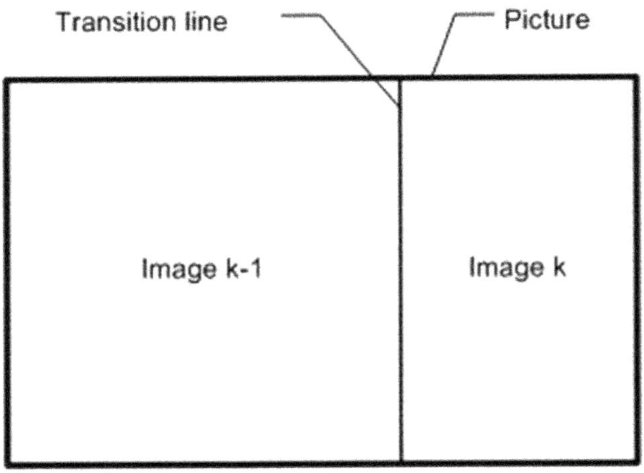

Figure 84: A picture displaying two sequence images.

In 3D lenticular pictures, successive images in the sequence are very similar to each other, and therefore it is difficult to notice the transition lines. Due to the finite resolution of the printer and the lenticular sheet, the transition line in real pictures becomes a transition region with a finite width, in which the display gradually changes from one image to the other. This gradient makes the transition lines even more difficult to spot. Practically, pictures displaying two or more successive images, like the one shown schematically in Figure 84, will look as if they present a single image. Although ideally, the picture should display a single image over its whole area, in 3D pictures a simultaneous display of two or more successive images is not considered to be a visual defect, and such a view is regarded as proper. This is not the case in general dynamic pictures, as will be discussed below.

There may be instances in which the picture will show a transition between non-sequential images. Such instance occurs in the image shown in Figure 84 if $k = 1$. In this case, the image "k-1" becomes image N, i.e., the last image in the sequence. Similarly, if k-1=N, then image "k" becomes image 1. In either case, the transition line will separate the first and the last images of the sequence. These images exhibit a considerable difference, and the transition line will be clearly seen and will cause a significant visual defect. Such a line is called a "break line," and a view displaying it will be considered as improper. This effect is illustrated in Figure 85 and Figure 86. A 24-image sequence of a 3D lenticular picture was chosen for this illustration.

Figure 85 shows a transition between images 1 and 2. A thin vertical line was superimposed on the transition line to denote it. Without this synthetic line, the transition line would be unnoticeable due to the similarity of images 1 and 2.

An improper view with a break line is illustrated in Figure 86. This break line is caused by a transition between images 1 and 24 and is clearly visible. In addition to the conspicuous visual defect, the break line also causes a stereoscopic inversion in the transition region. This amplifies the defect but can be illustrated only in a real 3D picture.

8.2 Viewing considerations of 3D lenticular pictures

The viewing space of a lenticular picture is divided into proper and improper viewing zones. In real 3D lenticular pictures, visual defects do not emerge abruptly as the observer crosses a proper viewing zone border; instead, they start to build up gradually as the observer approaches it. Therefore, the most pleasing view of the lenticular picture is obtained when the observer is located well within the proper viewing zone, preferably as far as possible from its borders.

Freedom of movement is one of the prime characteristics affecting the viewing comfort. Maximal movement freedom to both sides of the viewing point is highly desirable. This is another benefit of viewing lenticular pictures from the center of a viewing zone.

Pictures, in general, have a preferred viewing position. Whenever possible, the observers will position themselves next to the picture center. For pictures which are held in hand, the preferred viewing distance is about 50cm. For pictures hung on a wall, it is between x1.5 to x3 of the picture width. Central viewing is also characterized by a smaller maximal horizontal viewing angle. This is an additional benefit for viewing lenticular pictures since the optical resolution of lenticules decreases with the increase of the horizontal viewing angle. This is illustrated in Figure 87, where the viewing angles at the edges are denoted by α and β for the non-centered and the centered viewing respectively.

These considerations lead to the conclusion that lenticular pictures offer the best viewing experience (quality and comfort combined) when the preferred viewing position is at the center of a proper viewing zone. Due to manufacturing tolerances, there are variations in the geometry of the viewing zones. Minimization of the distance of the preferred viewing position from a proper viewing zone center is an important quality requirement for lenticular pictures.

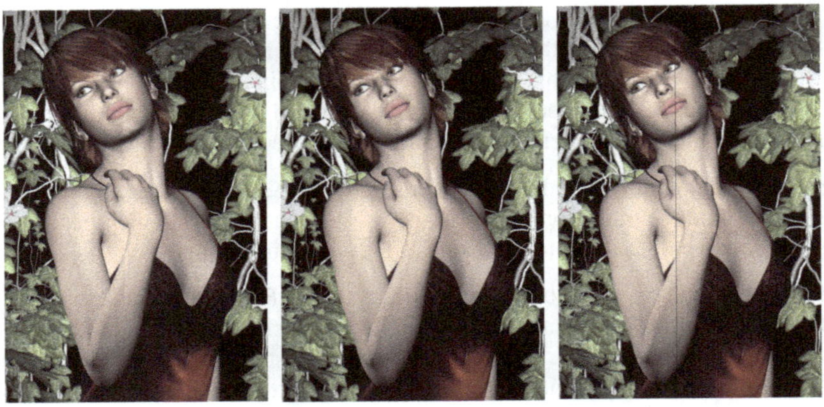

Image 1 Image 2 1-2 transition

Figure 85: Two successive images and an image displaying a transition between them. The thin vertical line denotes the transition line.

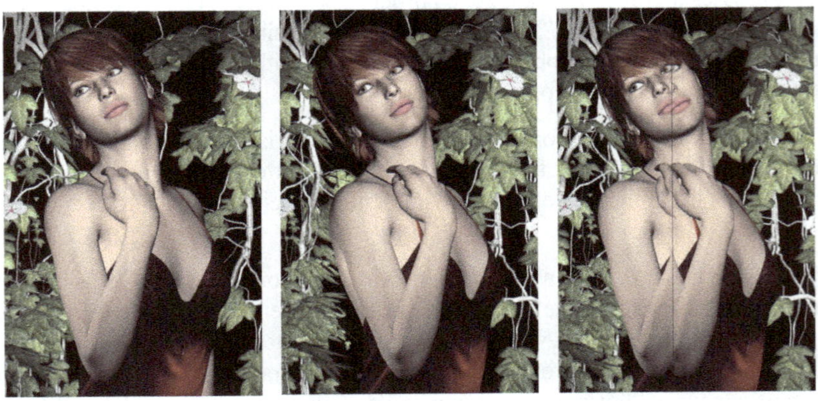

Image 1 Image 24 1-24 transition

Figure 86: Illustration of the break line in a 3D lenticular picture, showing the transition between image 1 and 24. The thin vertical line denotes the transition line, which is a break line in this case.

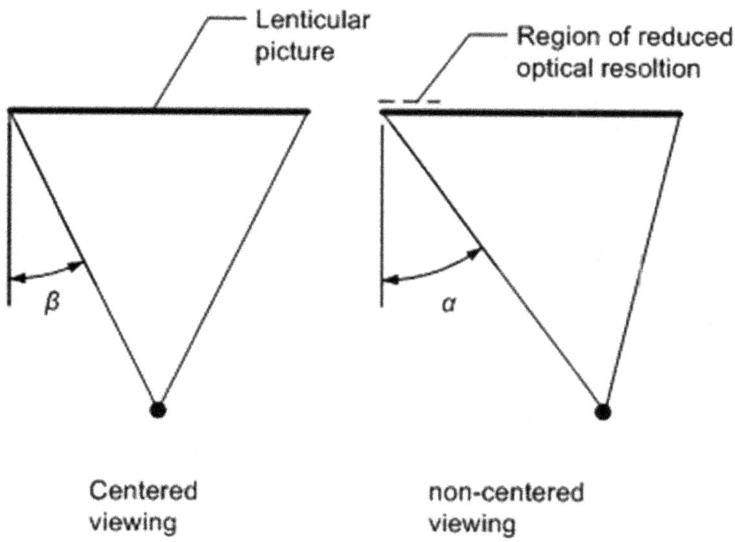

Figure 87: Centered versus non-centered viewing. The indication of the reduced optical resolution region is schematic and for illustration only.

8.3 The normals of a lenticular picture

Two geometrical objects are needed for the analysis: the central normal plane and the central normal. The central normal plane is a plane normal to the picture plane and passing through its vertical center-line. Central normals are all lines in this plane which are normal to the picture plane. The imaging planes are all planes which are normal to both the picture and the central normal planes. Each imaging plane contains a single central normal. All of these concepts are illustrated in Figure 88.

A coordinate system is defined in the cross-section plane, as shown in Figure 88. The V axis is also the central normal in its imaging plane.

The preferred viewing position is on the central normal plane.

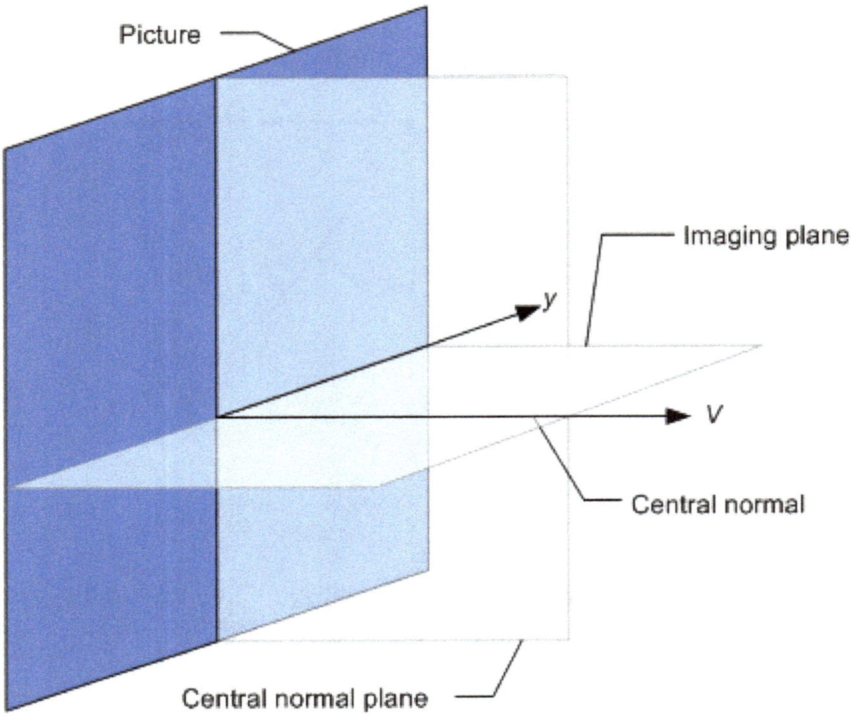

Figure 88: The lenticular picture and its normal.

8.4 Qualitative geometry of the proper viewing zones

The basic features of proper viewing will be derived by assuming that both the optical and the printed grids are perfectly straight and regular, and are perfectly aligned and parallel to each other. These are ideal conditions which are of course never met in practice. Some of the effects which arise from deviations from the ideal conditions assumed here will be discussed later.

The volume which allows proper viewing is a set of distinct volumes which are called “volume lobes." The intersections of the volume lobes with an imaging plane create a set of distinct areas, which are called "lobes." The lobes are shown schematically in Figure 89. In the ideal picture, the geometry of the lobes does not depend on the position (height) of the imaging plane.

As argued in paragraph 8.2, the best viewing experience is achieved when the preferred viewing position is located at the center of a lobe. This case is shown in Figure 89, and such a picture is regarded as being centered. The lobe containing the preferred viewing position is called the “main lobe," and all other lobes are called "secondary." There is an infinite number of secondary lobes, but due to the optical restrictions, only a few of them can be used for viewing. Figure 89 also shows the main viewing interval, introduced in paragraph 6.1.

All lobes are contained in a strip with edges parallel to the picture plane. The distances from the edges of this strip to the picture plane were denoted in Figure 89 by V_m and V_M respectively. The distance V_M can be infinite. At the calibration distance V_0 all adjacent lobes touch each other at a single point. Thus, viewing at the calibration distance is unobstructed laterally except at the singular meeting points.

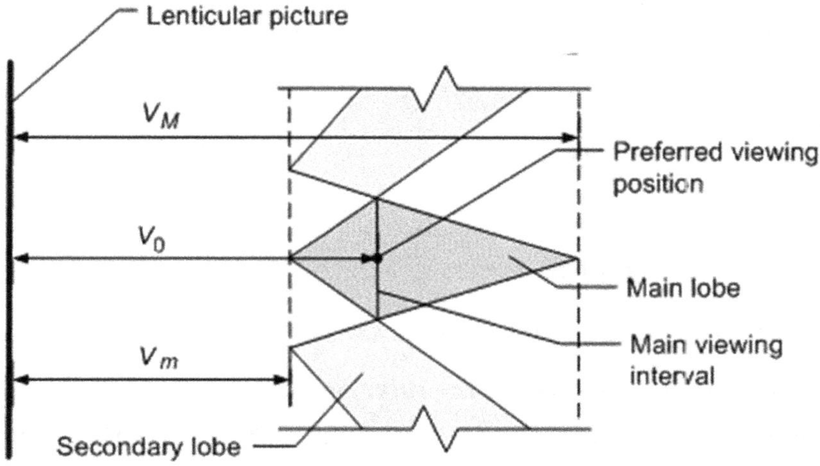

Figure 89: The lobes of a perfectly centered lenticular picture.

The observer will see the same object in all lobes. However, it will appear with the correct perspective only in the main lobe. Incorrect perspective causes a geometric distortion in the object display when it is viewed from secondary lobes.

Figure 89 shows an ideal situation in which the picture is perfectly centered. In real pictures, there is always a certain amount of de-centering, causing skewing of the lobes structure. The lobe structure of a slightly de-centered picture is shown schematically in Figure 90. This skewing causes the preferred viewing point to appear non-centered with respect to the main lobe, thus hindering the viewing experience. However, other properties of the lobes structure are preserved. In particular, de-centering does not affect the lobes strip boundaries V_m and V_M, and the laterally unobstructed viewing at V_0.

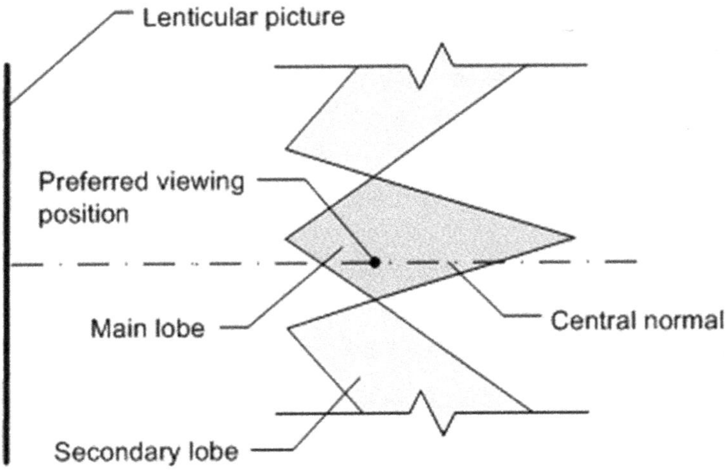

Figure 90: The lobes of a de-centered lenticular picture.

8.5 Coordinate systems for the interlaced picture

Let us consider a lenticular picture with N_L lenticules. As before, it will be assumed that N_L is odd.

First, a global coordinate X is defined for the horizontal direction of the interlaced picture. As explained in Chapter 4, this picture is composed of an array of cells, and the number of cells equals the number of lenticules. The origin of the global coordinate is defined as the center of the central cell. The global coordinates of the two extreme cells centers are shown in Figure 91. In this figure, the printed cells are represented as rectangles. The optical grid of this picture consists of N_L gridlines, one gridline per lenticule. Each optical gridline has a distinct global coordinate.

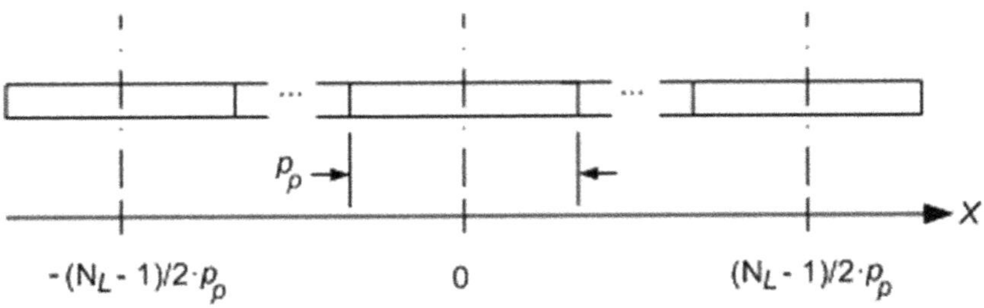

Figure 91: The global coordinate X.

Next, a set of local cell coordinates ξ_k is defined. The origin of a given local coordinate is defined to be the center of its cell. The central cell index is set to zero so that k takes values from $-(N_L\text{-}1)/2$ to $(N_L\text{-}1)/2$. Since there is a unique lenticule associated with each cell, and a unique optical gridline associated with each lenticule, the same index can be used to refer to a given lenticule and its optical gridline. In particular, the k'th optical gridline can be associated with the k'th local coordinate. In this manner, each optical gridline will be characterized by a pair of coordinates: the global and the local. The global and the local coordinates X_k and ξ_k of the k'th optical gridline are related by

$$X_k = kp_p + \xi_k. \qquad \text{Equation 73}$$

However, since adjacent optical gridlines are separated by p_o, the global coordinate can also be expressed as follows:

$$X_k = kp_0 + \xi_0. \qquad \text{Equation 74}$$

It now follows that

$$\xi_k = \xi_0 + k\Delta p, \qquad \text{Equation 75}$$

where Δp is the mismatch between the optical and the printed grids:

$$\Delta p = p_o - p_p. \qquad \text{Equation 76}$$

Note that all quantities in Equation 75 can be positive, negative or zero.

Alert readers will notice that optical gridlines were not directly associated with cells, only with lenticules. This is because in general, the k'th optical gridline may fall outside the k'th cell. Also, there may be cells with two optical gridlines or none.

There are three facts worth noting concerning the local coordinates.

1. When the picture is viewed from the calibration distance, the grids mismatch Δp vanishes, and therefore all local coordinates of the gridlines become equal,
2. For a centered picture $\xi_0 = 0$,
3. The local coordinates of all gridlines vanish for a centered picture when viewed from the calibration distance (this follows from 1 and 2 combined).

8.6 Quantitative analysis of lobes geometry: the main lobe of a perfectly centered picture

8.6.1 The content displayed by a lenticule

The content displayed by a given lenticule depends on the local coordinate of its optical gridline. Figure 92 shows an example of two cells having twelve sub-cells. In case (a), the optical gridline is located at the center of the first sub-cell, so its local coordinate is $-11/24 \cdot p_p$. The lenticule corresponding to this cell will display a sample of the first image in the sequence. In case (b), the optical gridline is at the center of the ninth sub-cell, so its local coordinate is $5/24 \cdot p_p$. The lenticule corresponding to this cell will display a sample of the ninth sequence image.

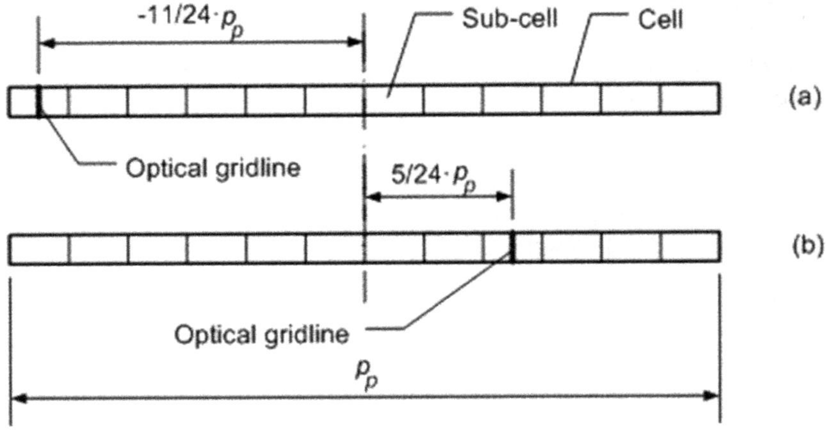

Figure 92: Two cells with optical gridlines having different local coordinates.

To formulate the general relation between the local coordinate and the index i (assumed to take values between 1 to N) of the displayed image it is necessary to distinguish between two cases, corresponding to even and uneven number of sub-cells. For the case of an even number of cells (like in the example shown in Figure 92), the relation is

$$i = \left\{ \frac{N\xi}{p_p} \right\} + \frac{N}{2}, \qquad \textit{Equation 77}$$

and for an uneven number of sub-cells it is

$$i = \left\{ \frac{N\xi}{p_p} + \frac{1}{2} \right\} + \frac{N-1}{2}, \qquad \textit{Equation 78}$$

where, in this case, $\{x\}$ is the ceiling function.

When the picture is viewed from the calibration distance, all optical gridlines have identical local coordinates. Therefore, all lenticules will display samples of the same image, and the whole picture will display a complete single image. However, when the printed and the optical grids do not match, the local coordinates of the optical gridlines will be different, and correspondingly different lenticules may display samples of different images.

8.6.2 The formation of a break line

A break line is formed when two adjacent lenticules display samples of images which are non-adjacent in the sequence. This happens when the local coordinate absolute value crosses $p_p/2$ (or, more generally, $(m + ½)p_p$). In the two lenticules associated with this crossing, one will display a sample of the first image in the sequence, and the other the last one. The formation of a break line is illustrated in Figure 93. In this example, each cell contains eight sub-cells. Furthermore, this example corresponds to the cases $\Delta p > 0$ and $k > 0$.

In cell k–1 the local coordinate of its gridline is close to $p_p/2$ and is contained in the 8'th sub-cell. Therefore, its corresponding lenticule will show a sample of the last image in the sequence (image 8). If the value of Δp is such that

$$\xi_k = \xi_{k-1} + \Delta p > \frac{p_p}{2},$$

the optical gridline corresponding to the k'th lenticule will migrate to the first sub-cell of the (k+1)'th cell so that it will display a sample of the first image in the sequence. There are no optical gridlines in cell k; instead of displaying a sample from a single image, this cell will display a mixture of samples from the 8'th and the first images. This mixed sample will act as a transition region between the displays of the 8'th and the first images samples in cells k-1 and k+1 respectively. Such transition will appear as a break line.

In Figure 93, cells k-1 and k+1 were surrounded by a broken line, to emphasize their boundaries. Cell k was not surrounded by a broken line for the sake of clarity.

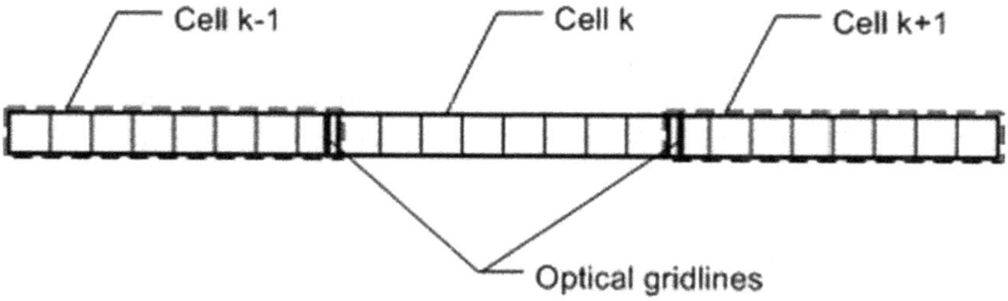

Figure 93: The formation of a break line for Δp > 0.

The formation of a break line for the case $\Delta p < 0$ and $k > 0$ is illustrated in Figure 94. It shows a cell with an optical gridline near its left edge so that its local coordinate value is slightly greater (algebraically) than $-p_p/2$. If the value of Δp is such that

$$\xi_k = \xi_{k-1} + \Delta p < -\frac{p_p}{2},$$

cell k-1 will display a sample of image 1, cell k+1 will display a sample of image 8, and cell k a mixture of samples from images 1 and 8. Although the geometry, in this case, is slightly different from the geometry displayed in Figure 93, the visual effect is the same: the k'th lenticule will display a break line.

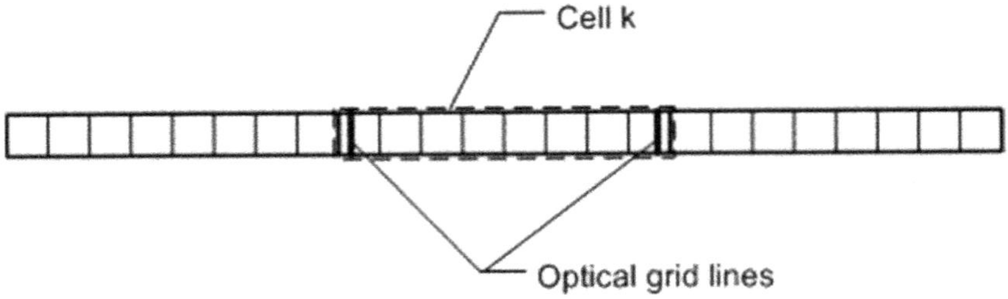

Figure 94: The formation of a break line for $\Delta p < 0$.

8.6.3 The optical grid displacement constraints for a centered picture

Lateral movement of the observation point causes displacement of the optical grid. Let us start by considering the maximal displacement of the optical grid which satisfies the proper viewing condition (no break line display) for a centered picture.

The optical gridlines that correspond to a viewpoint on a central normal are shown in Figure 95. This diagram shows three cells of the printed image: the central, the leftmost, and the rightmost. The broken lines denote the cell centers. The thicker lines denote the optical gridlines as seen from a particular viewing distance V.

Since it is assumed that this picture is centered, the optical gridline located in the central cell coincides with its center line. If the viewing distance is the calibration distance V_0, then the optical gridlines would be centered in all cells. For any other range, the optical gridlines will deviate from the cell centers in all but the central one. The deviation will increase towards the picture edges. Let $\xi^0{}_L$ and $\xi^0{}_R$ denote the local coordinates of the leftmost and the rightmost optical gridlines respectively. The upper part of Figure 95 corresponds to the case $V < V_0$ and the lower part for $V > V_0$.

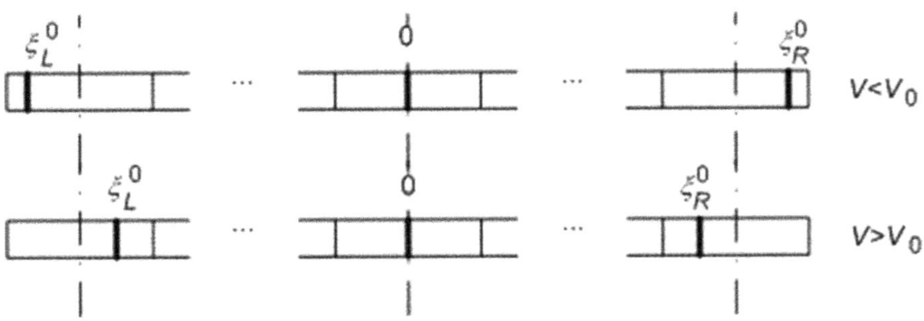

Figure 95: Centered optical grid at two viewing distances.

Let us introduce the quantity η:

$$\eta = \frac{N_L - 1}{2} \Delta p \ .$$

Using Equation 75 with $\xi_0 = 0$ yields

$$\xi_R{}^0 = -\xi_L{}^0 = \eta \,. \qquad \text{Equation 79}$$

When the observer moves their eye away from the central normal, the optical grid moves in the opposite direction. Figure 96 shows an optical grid shifted by ξ_0 with respect to the center.

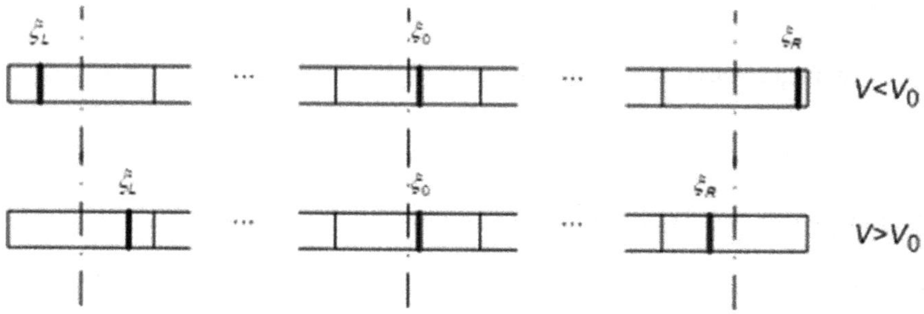

Figure 96: Shifted optical grid

Using Equation 75 again it is seen that

$$\xi_R = \xi_0 + \eta, \quad \xi_L = \xi_0 - \eta. \qquad \text{Equation 80}$$

For proper viewing, all optical gridlines should be contained within their corresponding cells. This condition can be formulated as follows:

$$|\xi_R|, \ |\xi_L| < \frac{p_p}{2}. \qquad \text{Equation 81}$$

Using Equation 80, these inequalities can be rewritten as follows:

$$-\frac{p_p}{2} - \eta < \xi_0 < \frac{p_p}{2} - \eta \,,$$

$$-\frac{p_p}{2} + \eta < \xi_0 < \frac{p_p}{2} + \eta \,.$$

Since both inequalities must be satisfied simultaneously, their right-hand sides can be shuffled so that they can be rewritten as

$$-\frac{p_p}{2} - \eta < \xi_0 < \frac{p_p}{2} + \eta$$

and

$$-\frac{p_p}{2} + \eta < \xi_0 < \frac{p_p}{2} - \eta \,,$$

which is equivalent to

$$|\xi_0| < \frac{p_p}{2} + \eta,$$

and

$$|\xi_0| < \frac{p_p}{2} - \eta.$$

Both these inequalities can be combined to a single inequality

$$|\xi_0| < \frac{p_p}{2} - |\eta|. \quad \textit{Equation 82}$$

From Equation 12 and Equation 17 it follows that

$$\Delta p = \frac{pt}{n}\frac{V_0 - V}{V_0 V}. \quad \textit{Equation 83}$$

This relation reveals the dependence of ξ_0 on the viewing distance. From Equation 17, it follows that

$$|\eta| = \frac{N_L - 1}{2}\frac{pt}{n}\frac{|V_0 - V|}{V_0 V},$$

so that finally

$$|\xi_0| < \frac{p_p}{2} - \frac{N_L - 1}{2}\frac{pt}{n}\frac{|V_0 - V|}{V_0 V}. \quad \textit{Equation 84}$$

This condition gives the maximal allowed optical grid shift amplitude as a function of the viewing distance.

8.6.4 Relation between displacements

A displacement ε of the optical grid is caused by a specific displacement E of the observation point. The relation between these displacements is described in Figure 97. Using triangular similarity relation, one gets

$$E = -\frac{nV}{t}\varepsilon. \quad \textit{Equation 85}$$

In this equation, the displacements E and ε are regarded as signed quantities, with the sign being determined by the displacement direction. The minus sign in Equation 85 reflects the fact that the two displacements are in opposite directions. The coordinates of the points E_L and E_R corresponding to the left and right boundaries of the main lobe can be now written as:

$$E_L = -\frac{nV}{2t}\left(p_p - \frac{(N_L-1)pt}{n}\frac{|V_0 - V|}{V_0 V}\right),$$ Equation 86

and

$$E_R = \frac{nV}{2t}\left(p_p - \frac{(N_L-1)pt}{n}\frac{|V_0 - V|}{V_0 V}\right).$$ Equation 87

8.6.5 *The main lobe for a centered picture*

Let us consider the trajectory of the point E_R in the imaging plane. For this we need to separate Equation 87 into two cases: $V < V_0$ and $V > V_0$. For the case $V < V_0$ one gets:

$$E_R^{<} = \frac{nV}{2t}\left(p_p - \frac{(N_L-1)pt}{n}\frac{V_0 - V}{V_0 V}\right) \approx \left(\frac{R_p}{2} + \frac{W}{2V_0}\right)V - \frac{W}{2}$$ Equation 88

where W is the picture width. Similarly, for the case $V > V_0$ one gets the expression

$$E_R^{>} = \frac{nV}{2t}\left(p_p + \frac{(N_L-1)pt}{n}\frac{V_0 - V}{V_0 V}\right) \approx \left(\frac{R_p}{2} - \frac{W}{2V_0}\right)V + \frac{W}{2}.$$ Equation 89

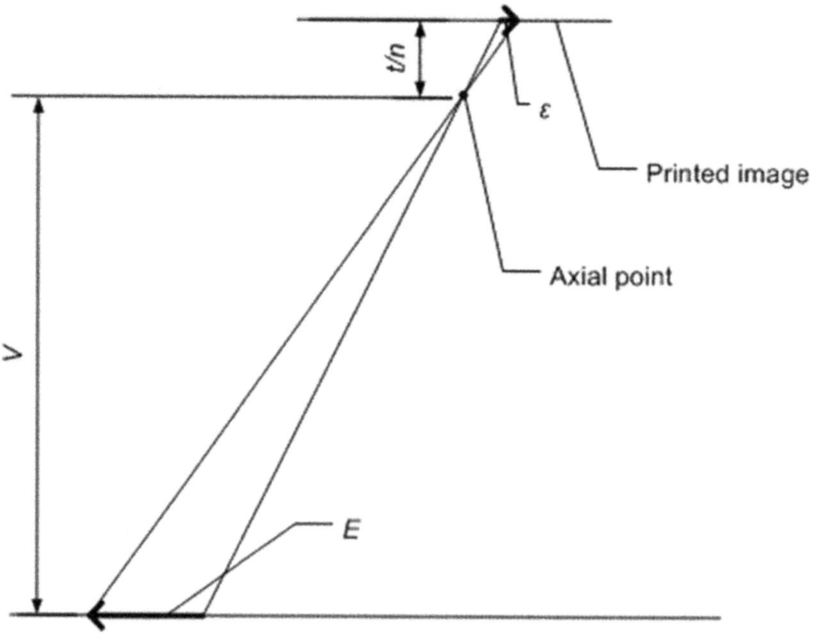

Figure 97: The relation between optical grid and viewing point displacements.

In the second step of Equation 88 and Equation 89 the following approximations were used:

$$W = (N_L - 1)p\,,$$

and

$$p_p \sim p.$$

Both $E_R^<$ and $E_R^>$ are linear in V, so Equation 88 and Equation 89 represent straight lines in the imaging plane. For $V = 0$, $E_R^< = -W/2$ and $E_R^> = W/2$, and for $V = V_0$, $E_R^< = E_R^> = R_p V_0/2$. These relations define the lines.

The lines $E_R^<$ and $E_R^>$ define the right edge of the proper viewing interval. The left edge is determined by the analogous $E_L^<$ and $E_L^>$ which are derived from Equation 86. These lines are reflections of $E_R^<$ and $E_R^>$ with respect to the central normal.

There are two cases regarding the geometry of the lobes:

1. Finite longitudinal range, $R_p < W/V_0$
2. Infinite longitudinal range, $R_p > W/V_0$

The singular case $R_p = W/V_0$ marks the transition between the two cases. It represents an infinite longitudinal range with a laterally bounded lobe.

The geometries of the main lobe are shown in Figure 98, Figure 99, and Figure 100. The left edge of the lobe is the reflection of the right edge with respect to the central normal.

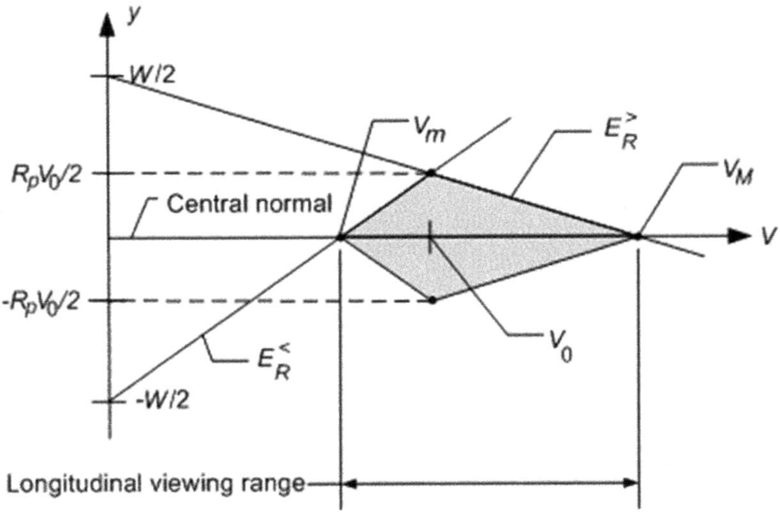

Figure 98: Lobe geometry for $R_p < W/V_0$ (Case 1).

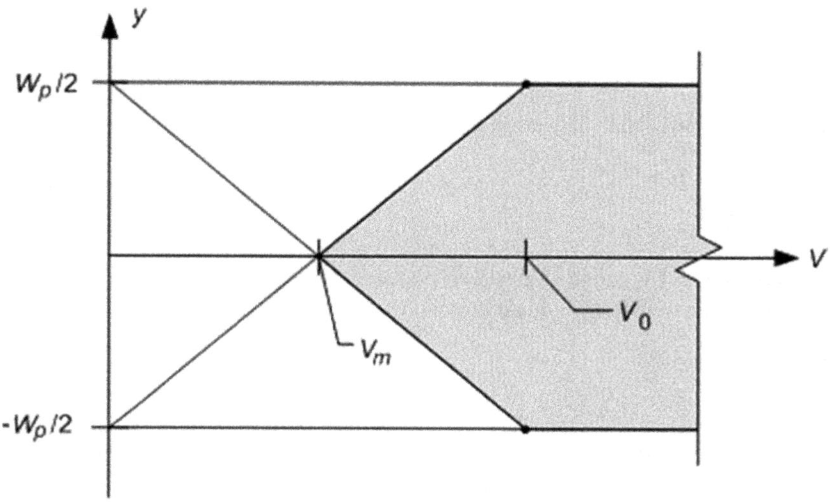

Figure 99: Lobe geometry for the singular case $R_p = W/V_0$.

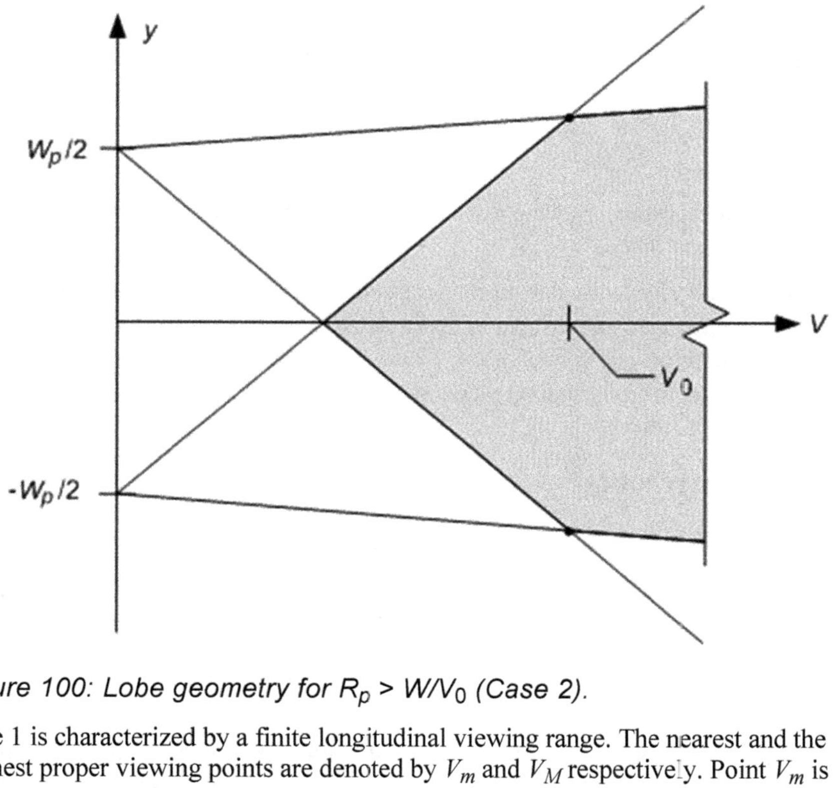

Figure 100: Lobe geometry for $R_p > W/V_0$ (Case 2).

Case 1 is characterized by a finite longitudinal viewing range. The nearest and the furthest proper viewing points are denoted by V_m and V_M respectively. Point V_m is derived by setting $E_R^<$ to zero:

$$V_m = \frac{V_0 W}{W + R_p V_0} \; . \qquad \textit{Equation 90}$$

Likewise, V_M is derived by equating $E_R^>$ to zero:

$$V_M = \frac{V_0 W}{W - R_p V_0} \; . \qquad \textit{Equation 91}$$

Quite often the artist is asked to design a picture which will be viewable from a large distance. To achieve this, they must choose V_0 such that

$$V_0 \geq \frac{W}{R_p} \, .$$

However, to allow proper viewing at the shortest distance, the smallest value of V_0 can be chosen:

$$V_0^s = \frac{W}{R_p} \; . \qquad \textit{Equation 92}$$

Substituting this value for V_0 in Equation 90 gives

$$V_m^s = \frac{W}{2R_p} \; .$$

This is the shortest proper viewing distance which can be achieved while maintaining proper viewing at infinity.

The value of R_p for lenticular sheets used to make 3D lenticular pictures is around 0.5, and the typical viewing distance is about twice the image width (but, of course, at least 30cm). Therefore, for 3D lenticular pictures $R_p \sim W/V_0$, which is near the singular case. Accordingly, the longitudinal viewing range of 3D lenticular pictures is typically large, and often infinite.

8.7 Off-center lobes

8.7.1 General

In paragraph 8.6, the boundary contour of the main lobe was derived for a perfectly centered picture. A lenticular picture contains secondary lobes which appear off-center. Also, in non-centered pictures, the main lobe will appear off-center as well. Since a real lenticular picture is never perfectly centered, we may say that in practice all lobes are off-center.

Fortunately, the boundary contour of any off-center lobe can be derived from the boundary contour of the centered lobe. So, instead of analyzing the non-centered case in general, it is shown here how to modify the centered geometry to account for the off-center position. This method is valid for both main and secondary and lobes, in both centered and non-centered pictures.

8.7.2 *The non-centered main lobe*

As already illustrated in Figure 90, the main lobe will appear off-center if the picture is non-centered. A picture is non-centered when the optical grid corresponding to a central normal view has a non-zero central local coordinate.

Let us reconsider Figure 96, which shows a shifted optical grid with non-zero central local coordinate ξ_0. In the context of paragraphs 8.5 and 8.6, this shift was caused by a shift in observation from the central normal. Here, this shift is the result of picture de-centering.

A given shift of the optical grid causes a shift in the viewing space according to Equation 85. The lobe geometry will be modified by adding a line $-n\xi_0 V/t$ to the contour of the centered picture lobe. This transformation will cause skewing, as illustrated in Figure 101, and has the following properties:

1. All shifts are parallel to the y axis,
2. For a given viewing distance V all shifts are of the same magnitude and direction and equal to $-n\xi_0 V/t$.

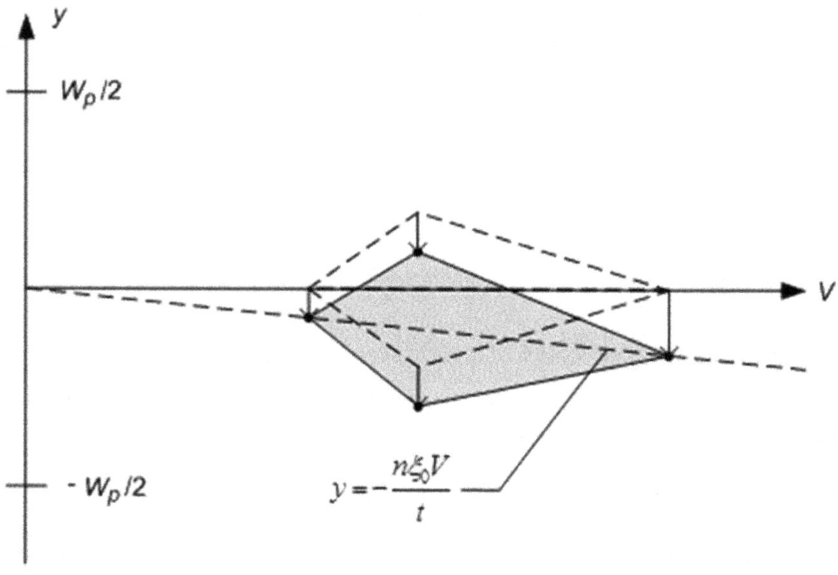

Figure 101: The main lobe in a non-centered picture.

Due to property 1, the skewing transformation does not affect the longitudinal viewing interval, and due to property 2, the lateral viewing interval size is not affected. Both remain the same as in the centered case.

Figure 101 shows the case of a lobe with finite longitudinal range. It is left to the reader to construct the off-center main viewing lobe for the other two cases corresponding to Figure 99 and Figure 100.

8.7.3 Secondary lobes in a centered picture

The boundary contours of secondary lobes adjacent to the main lobe in a centered picture are derived by de-centering the picture by a single cell width p_p. Applying the corresponding skewing transformations to Figure 98, the scheme shown in Figure 102 is obtained. The top and bottom secondary lobes were derived by shifting the optical grid by $-p_p$ and $+p_p$ respectively.

The boundary contours of the next pair of secondary lobes are derived from the boundary contour of the main lobe by introducing de-centering of $\pm 2p_p$, etc. Thus, the whole lobe geometry can be easily derived from the main lobe geometry.

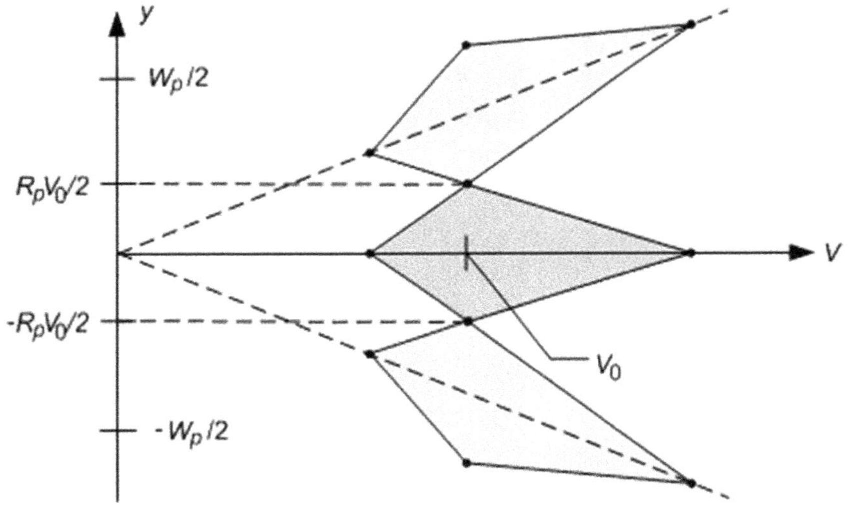

Figure 102: Secondary lobes of a centered picture

Again, the scheme in Figure 102 is for the finite range case. The reader is invited to derive the corresponding schemes for the two other lobe geometry case.

8.7.4 Secondary lobes in non-centered picture

The boundary contours of the secondary lobes of a non-centered picture can be derived by the same method as outlined in 8.7.3 for the main lobe shown in Figure 101. The result will be a skewed version of the scheme shown in Figure 102. The unobstructed proper viewing at V_0 is not affected by picture de-centering.

8.8 The effect of binocular vision

So far, the discussion was limited to monocular vision. Observing the picture with both eyes has a significant impact on proper viewing. In particular, it reduces the extent of the proper viewing lobes and introduces gaps between them. Unobstructed lateral viewing is impossible with binocular vision.

The effect of binocular viewing is illustrated in Figure 103. This figure shows two lobes: the main one and a secondary one adjacent to it. The observer's pupils are represented by black dots, and the center point between them is denoted by a cross. This location is sometimes referred to as the position of the "cyclopean" or "third" eye.

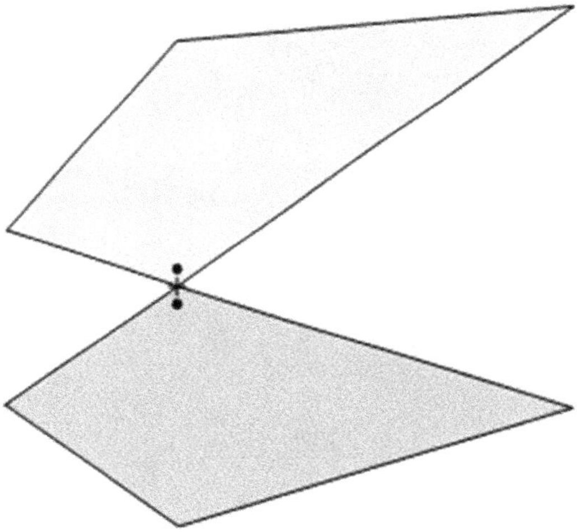

Figure 103: Binocular viewing in monocular lobes.

For monocular viewing, all the shaded area allows proper viewing. Figure 103 shows an observer with their third eye located near the tangent point. In this position the observer's eyes are in different lobes, and they will be exposed to reversed projections of the object. This will cause a pseudoscopic inversion of depth, which is a severe visual defect. Such a view is regarded as improper, and this position must be excluded from the proper viewing region. Therefore, binocular viewing introduces an additional requirement for proper viewing: both eyes of the observer must be in the same lobe. This new condition reduces the lobe area. The area within the lobe which allows proper viewing with both eyes will be called the "binocular lobe."

Figure 104 shows the main binocular lobe and the two secondary binocular lobes adjacent to it. The binocular lobes are derived from their respective lobes by placing the observer at the four corners of the monocular lobe and marking the positions of his third eye. This gives four points which constitute the new corners of the binocular vision lobe. Figure 104 shows both the original lobes and their respective binocular lobes for comparison. The shaded regions denote the original lobes, and the white areas, surrounded by the broken line, the binocular lobes.

This procedure creates gaps between the binocular lobes, as illustrated in Figure 104. These gaps are called "binocular gaps". In contrast to monocular viewing, they prohibit unobstructed lateral viewing at the calibration distance. Moreover, the width of these gaps is the inter-ocular distance (approximately 65mm), regardless of the lenticular picture parameters.

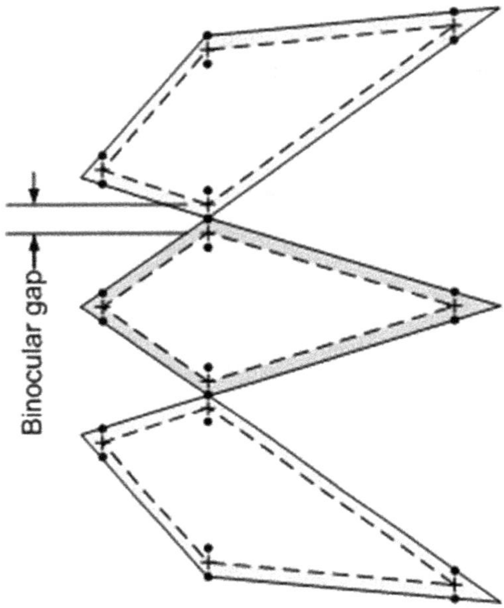

Figure 104: Proper viewing regions for binocular vision.

8.9 Depth display for general viewing distances

8.9.1 General

The depth display characteristics of a 3D lenticular picture were analyzed in Chapter 6 for viewing from the calibration distance V_0, where it was established that this distance does not affect the displayed depth. However, it is well known that in 3D lenticular pictures this is true, at least approximately, for a general viewing distance, not just V_0.

The independence of the imaged depth from the viewing distance is often explained by using a heuristic argument. It may be said that a 3D lenticular picture shows approximately the correct projection of the object for any given horizontal viewing angle, regardless of the viewing distance. Therefore, each eye of the observer will see the correct projection for any viewing distance. This argument is correct for viewing at the calibration distance but becomes inaccurate for distances in which the picture displays more than one sequence image over its aperture, as shown schematically in Figure 84. A more detailed analysis is required to elucidate the depth display in such case. This analysis is presented below, and its conclusion supports the common experience. However, the presence of transition lines can interfere with the depth display, slightly modify its magnitude and cause small depth undulations. Under certain conditions, these undulations are clearly visible.

The depth display invariance with respect to the viewing distance enhances the viewing experience of lenticular pictures significantly. The observer can enjoy a correct 3D view while moving in both parallel and perpendicular directions with respect to the picture plane, as long as they remain in the same lobe. If the observation is from the main lobe, the object will also appear without geometrical distortions. This is in sharp contrast to stereoscopic displays based on only two images. In such displays, the visual parallax is fixed and does not depend on the viewing distance. This causes the displayed depth to increase with the viewing distance (Mendiburu, 2009), thus introducing severe distortions for movement perpendicular to the picture plane. Stereoscopic pictures also exhibit distortions for parallel movement, since they can display only one projection per eye.

8.9.2 The effect of transition lines in the visual parallax

When the observation point is shifted from one eye of the observer to the other, the optical grid is displaced accordingly. Denoting by ξ_0 the local coordinate of the central optical gridline corresponding to the location of the "third eye," the local coordinates of this gridline for the two eyes are given by (Equation 85)

$$\xi_0{}^L = \xi_0 + \frac{te}{2nV}, \quad \xi_0{}^R = \xi_0 - \frac{te}{2nV},$$

where e is the interocular distance. This shift causes a change in the picture display. Let us focus on a particular lenticule m, and estimate the change of the indices of the images it displays when the optical grid is shifted in this manner. The local coordinates corresponding to the left and the right views for the m'th lenticule are derived from Equation 75:

$$\xi_m{}^L = \xi_0 + \frac{te}{2nV} + m\Delta p, \quad \xi_m{}^R = \xi_0 - \frac{te}{2nV} + m\Delta p\,.$$

The corresponding indices of the displayed images in lenticule m can be derived from Equation 77 or Equation 78. For the sake of simplicity, we ignore here the discrete nature of the image index and treat it as a continuous number. Within this approximation, one obtains

$$i_m{}^L = \frac{N\xi_m{}^L}{p_p} + \frac{N}{2}, \quad i_m{}^R = \frac{N\xi_m{}^R}{p_p} + \frac{N}{2}, \qquad \text{Equation 93}$$

so, the index difference is

$$\Delta i_m = i_m{}^L - i_m{}^R = \frac{Nte}{nVp_p} \approx \frac{N}{R_p}\frac{e}{V}, \qquad \text{Equation 94}$$

where in the last step p_p was approximated by p.

Let us assume that the interval $[\xi_m^L, \xi_m^R]$ does not include a cell boundary. In this case, the visual parallax for any given imaged object point between images i_m^L and i_m^R will be given by

$$P_v^m = \Delta i_m P_s = \frac{P}{R_p}\frac{e}{V}, \qquad \text{Equation 95}$$

where P_s is the corresponding incremental parallax. This is identical to Equation 47, except that V_0 is replaced by a general viewing distance V. However, the real visual parallax may be different from P_v^m due to interference caused by the presence of transition lines. The reason for this discrepancy is that P_v^m was computed for a given lenticule m, whereas the left and right views representations of a given object point will be separated by the true visual parallax P_v, and will appear in different lenticules.

The difference between P_v^m and P_v can be analyzed with the help of Figure 105. This figure shows a certain object point whose representations are annotated by A and B for the right and the left views respectively, each being displayed by a different lenticule. Since point A appears to the left of point B, the corresponding imaged point is protruding. Let us assign the index m arbitrarily to the lenticule displaying point A. The figure shows two cases. In case (a) there is no transition line between points A and B in the right view. In case (b) there is a transition line between points A and B in the right view. The transition lines in the left view are not shown since their locations are irrelevant.

The distance between points A and B is the true visual parallax P_v. In case (a), the image index displaying point A is i_m^L, and the image index displaying point B is i_m^R. The index difference between the images displaying points A and B is given in this case by Equation 94, and the true visual parallax is equal to P_v^m as given by Equation 95. On the other hand, in case (b) there is a transition line between points A and B, causing the image index displaying point B to be different from i_m^R by unity. Such transition line will be regarded as "interfering." In the example shown in Figure 105 (b), the image index of point B is $i_m^R - 1$, but, in general, it can also be $i_m^R + 1$, depending on whether $V < V_0$ or $V > V_0$. This causes a discrepancy of magnitude P_s between P_v^m and P_v.

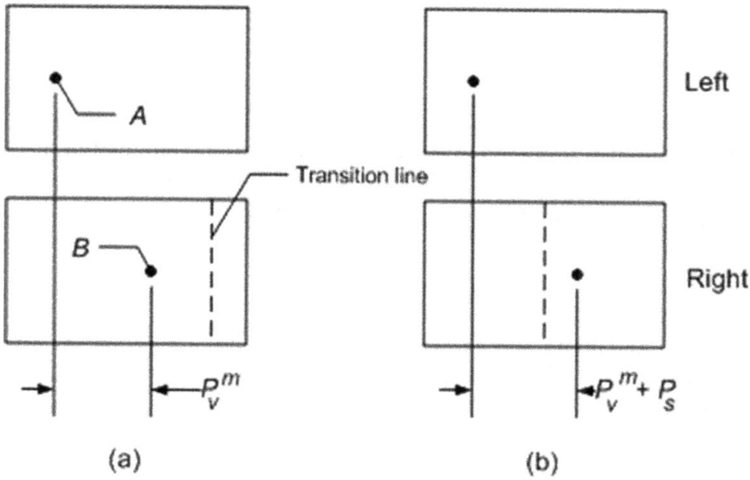

Figure 105: Interfering transition line.

8.9.3 *Depth display in the absence of interfering lines*

If $P_v^m = P_v$, then the analysis shown in paragraph 6.4 is valid for any viewing distance, not just V_0. As noted already, Equation 47 is valid in this case for any viewing distance V. After replacing V_0 by V in Figure 53 it can be easily shown that Equation 48 through Equation 53 are valid for a general viewing distance V. The small parallax approximation condition becomes $h << V$. In the absence of transition lines interference, Equation 54 is valid for all viewing distances satisfying this condition. This demonstrates the validity of the depth display invariance with respect to the viewing distance in the absence of interfering lines.

8.9.4 *Depth display in the presence of interfering lines*

Changes in the visual parallax induce changes in the displayed depth. These changes occur near the transition lines. Since the positions of the transition lines change with the movement of the observer, their interference causes undulations in the depth display which move with their motion. In stationary viewing, the depth undulations are unnoticeable. However, when the observer moves, the undulations move too and often become noticeable.

Let us denote the imaged depth when affected by a single transition line interference by h_t. Its value can be derived from Equation 51:

$$h_t = \frac{V}{e} P_v = \frac{P}{R_p} \pm \frac{V}{e} P_s,$$

thus, the magnitude of the depth undulations u_t is

$$u_t = \frac{P_s}{e} V .$$

For a given V, the maximal undulation magnitude is obtained for points with maximal P_s, namely, for points imaged at the largest distance from the picture plane. Let us assume that the whole picture is displayed continuously. In this case, the maximal value of P_s is p, so the maximal undulation magnitude becomes

$$u_{t,\max} = \frac{p}{e} V .$$

It is seen that the maximal undulation amplitude is proportional to the lenticule pitch p; thus, the undulations visibility increases with decreasing lenticules density. Let us consider an example of 20lpi lens viewed at 1m. In this case

$$u_{t,\max} \approx 2\text{cm} .$$

In general, there may be more than one line interfering with the depth display. If there are N_t interfering lines, the imaged depth will be

$$h_t = \frac{V}{e} P_v = \frac{P}{R_p} \pm N_t \frac{V}{e} P_s . \qquad \textit{Equation 96}$$

Let us consider an observer at a distance V watching an object point with N_t interference lines. When this observer moves horizontally and parallel to the picture plane, the number of interfering lines will fluctuate between N_t and $N_t + 1$, or between N_t and $N_t - 1$, depending on the circumstances. So, regardless of the value of N_t, the undulations amplitude remains the same. However, the average value of the depth display is affected. Let us derive an estimate for N_t.

The indices k_q of the lenticules exhibiting transition lines can be derived from Equation 75, noting that in these lenticules the local coordinate is on a sub-cell boundary:

$$\xi_0 + k_q \Delta p = \left(q + \frac{1}{2} \right) \frac{p_p}{N} ,$$

where q denotes the serial number of the transition line. The index difference Δk between lenticules showing successive transition lines is

$$\Delta k = k_{q+1} - k_q = \frac{p_p}{N \Delta p} .$$

Let us assume again that the picture is continuous, so the maximal visual parallax is Np. In such case, there will be viewing positions with N_t interfering transition lines. The number N_t is given by (again ignoring the fact that N_t is an integer)

$$N_t = \left|\frac{P_{v,\max}}{\Delta kp}\right| = \frac{Np}{p}\frac{N|\Delta p|}{p_p} \approx N^2\frac{|\Delta p|}{p},$$ *Equation 97*

where in the last step p_p was approximated by p. Using Equation 83, this can be rewritten as

$$N_t = N\frac{h'}{V_0}\frac{|V_0 - V|}{V},$$ *Equation 98*

where h' is the limit of continuous imaging (Equation 56). When observing the picture from an arbitrary point, N_t can be arbitrarily large. However, it is of interest to determine what are the maximal values of N_t for observation from within a proper viewing zone.

It was shown that the range of the proper viewing zone (for monocular viewing) extends from V_m to V_M (Equation 90 and Equation 91). Substituting V_m or V_M for V in Equation 98, one obtains

$$N_t^p = NR_p\frac{h'}{W}.$$

For an infinite proper viewing range, the value of N_t^p decreases. Therefore, it can be concluded that N_t^p is the maximal value of N_t which can occur from within a proper viewing zone. The value of h'/W is typically of the order of 0.1, and for 3D pictures $R_p \sim 0.5$, so that

$$N_t^p \approx 0.05N.$$

This estimate demonstrates that to experience more than one interference line, the sequence length must be larger than ~20. Note that there will always be depth undulations for any viewing distance different from V_0, even if the estimate of N_t (Equation 97) gives a value which is less than 1.

8.10 The effect of grids misalignment

Until now it was assumed that the printed and the optical grids are perfectly parallel. Due to manufacturing tolerances, there is always a certain degree of misalignment between these grids. Such misalignment causes a reduction in the proper viewing area.

Figure 106 shows the central printed cell rectangle. For the sake of clarity, its aspect ratio (width/height) was superficially significantly increased. The center optical gridline is shown by a broken curve. For convenience it was assumed that the optical gridline passes through the cell's center.

The misalignment shown in Figure 106 causes a linear variation of the central local coordinate along the height of the picture. Its value at the top is denoted by a. Accordingly, its value at the center is 0, and at the bottom $-a$.

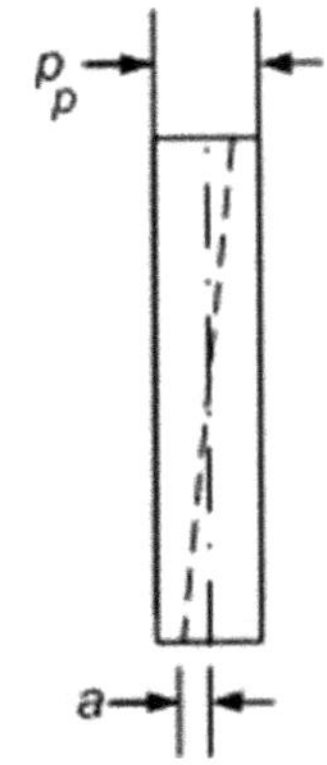

Figure 106: Misaligned grids.

In such a picture, the lobes structure will depend on the vertical angle of the line of sight. The lobes corresponding to an observation of the picture bottom will correspond to de-centering of $-a$, and the lobes corresponding to observation of the picture top will correspond to a shift of $+a$. These lobes will continuously transform from one to the other when the observer moves his observation point from one extreme position (top or bottom) to the other. The lobes corresponding to the extreme vertical viewing angles are shown in Figure 107.

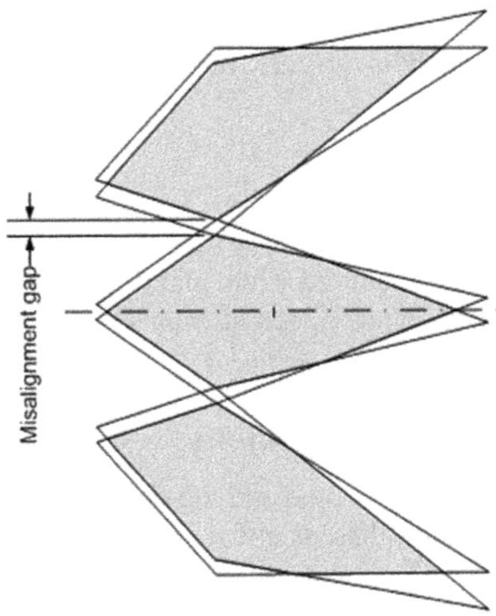

Figure 107: Lobes in a picture with misaligned grids.

The two most divergent lobes correspond to the top and to the bottom of the picture, and their overlapping areas are shown in gray color in Figure 107. It is evident that misalignment introduces gaps between the lobes and reduces both the transverse and the longitudinal proper viewing ranges.

As the observer moves their observation point from the bottom to the top of the picture, the lobes evolve, and their areas slightly change. The viewing lobes corresponding to the whole picture will be the common areas of all these lobes. The boundaries of these areas will not be straight lines, as in the ideally aligned case, but they will look similar to the common areas shown in Figure 107. The break line in a misaligned picture will be inclined, instead of being vertical as shown in Figure 86.

The aspect ratio of a cell is the aspect ratio of the picture divided by the number of lenticules. The angle corresponding to the cell aspect ratio in radians is the misalignment limit. A reasonable requirement is that the misalignment should be less than 0.1 of this angle value. Considering a square picture with 1000 lenticules, this sets a misalignment tolerance of 100μrad, which is about 20 arcseconds.

8.11 Dynamic pictures

The requirements of proper viewing in dynamic pictures are even more strict than with 3D pictures. Since in general, the sequence images in a dynamic picture can be completely different from each other, for proper viewing the lenticular picture must display a single sequence image. Any transition line in a dynamic picture is regarded as a break line.

The analysis of the proper viewing zones in dynamic pictures is very similar to the analysis of the proper viewing zones in 3D pictures which was presented in paragraph 8.6. All that is needed is to replace p_p by the sub-cell size c in Equation 81, which will read as

$$|\xi_R|,\ |\xi_L| < \frac{c}{2}\ ,$$

and then follow all of the subsequent development.

Figure 108 shows the lobe structure of a dynamic picture with a three-image sequence, in which $c = p_p/3$. The main lobe of a 3D picture made with the same parameters and characteristics is also shown for comparison. This main lobe is broken up into three sub-lobes, one sub-lobe per image.

The first striking fact evident from Figure 108 is that the sub-lobes of the dynamic picture are much smaller compared to a 3D picture. Laterally, all three sub-lobes of the dynamic picture fit within the main lobe of the 3D picture. The longitudinal viewing distance range of the dynamic picture is much smaller. Thus, proper viewing of dynamic pictures is much more constrained and less comfortable compared to 3D pictures. The differences become even more pronounced when the number of dynamic picture images increases.

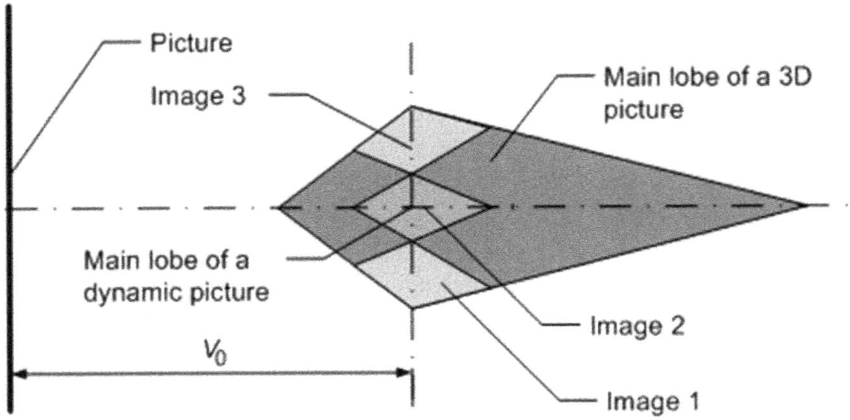

Figure 108: Comparison of lobes of dynamic and 3D pictures for the same picture size, calibration distance, and lens.

The images in a dynamic picture are displayed periodically. When the observer moves from a given lobe to the adjacent one, the image will switch to the adjacent image in the sequence. As the observer continues their movement passing from one lobe to the other all images in the sequence can be seen. Once all images are exhausted, they will be displayed sequentially again as the observer continues with his movement to the secondary sub-lobes (not shown in Figure 108).

8.12 The lenticular picture as a stereoscopic viewer

Making a dynamic picture from a stereoscopic image pair is possible. Such picture may be called "stereoscopic", and its lobes are shown schematically in Figure 109. In this application, the lenticular picture serves as a stereo viewer. Note that in this case (as in all centered dynamic pictures with an even number of images), the preferred viewing position is located precisely at the tangent point of two sub-lobes.

For proper stereoscopic viewing the left sub-lobe should correspond to the left eye image of the stereoscopic pair, and the right sub-lobe to the right eye image. This is illustrated in Figure 109. The viewing region of a stereoscopic picture is very small. It is composed of all third-eye locations which allow the observer to see a different picture in each eye. The center of this region is the tangent point of the main lobes pair.

This region can be determined by placing the viewer in four extreme locations around this point, as shown in Figure 110. As in Figure 103 and Figure 104 the observer's pupils are represented by dots and the third eye by a cross. The four third eye positions mark the apices of the viewing region, shown hatched. Its width is equal to the interocular separation.

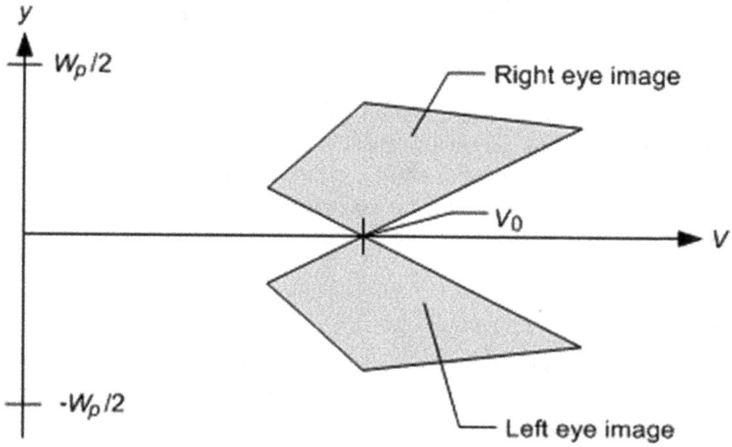

Figure 109: Lenticular picture as a stereoscopic viewer

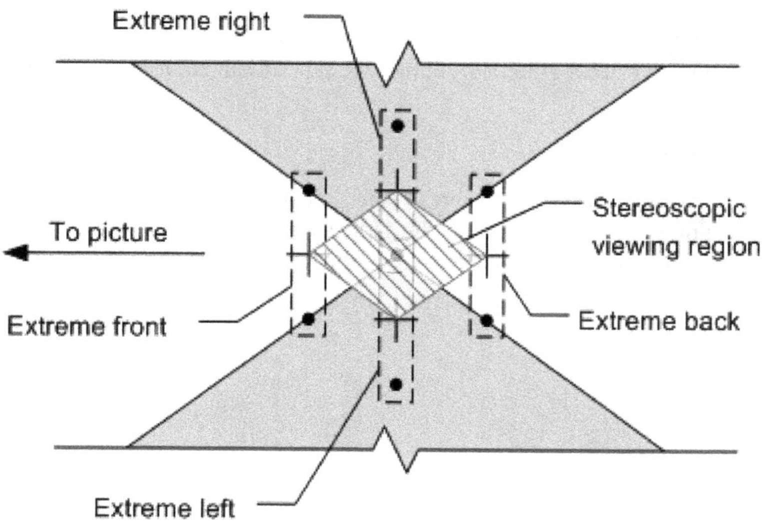

Figure 110: Viewing region for a stereo lenticular picture.

In spite the small size of the viewing region, most viewers can enjoy lenticular stereoscopic pictures if the center of the viewing region is at the preferred viewing position. Also, the printed and the optical grids must be parallel to a very high degree, to avoid the misalignment gap shown in Figure 107. Such a gap will further reduce the viewing region and may render the picture unviewable.

These requirements pose very tight tolerances on the manufacturing of stereoscopic lenticular pictures. These tolerances become more severe as the size of the picture increases.

8.13 Other mechanisms which affect the viewing regions

8.13.1 Resolution

Since lenticular pictures have finite resolution, the lobes boundaries are not sharp and possess a finite width. Thus, an observer will start to see a break line as they approach the boundary of the lobe. This has a similar effect as binocular viewing; a certain region near the lobe boundaries must be excluded from the proper viewing regions. The width of this region depends on the printer resolution and the optical parameters of the lenticular sheet.

8.13.2 Distortions

In practice both the optical and the printed grids are not ideal. A quantitative optical analysis (Chapter 11) reveals that the optical grid is slightly curved even if the lenticular sheet is perfect. Real lenticular sheets have distortions, which cause additional distortions in the optical grid. Indeed, real printers have distortions too. Thus, discrepancies between the printed and the optical grids exist not only due to misalignment but also due to distortions.

One can expect that the magnitude of discrepancy between the grids will increase with the picture size. This effectively limits the size of lenticular pictures which can be made.

8.13.3 Slanting

In slanted displays (Chapter 7), the break line appears slanted. The effect of this phenomenon on the viewing region is the same as in a misaligned picture (paragraph 8.9). Thus, slanted displays have reduced viewing regions.

Chapter 9
Content preparation for a 3D lenticular picture

"Any man who is truly worthy must consider himself unworthy."

- Robert Harris, in "Conclave"

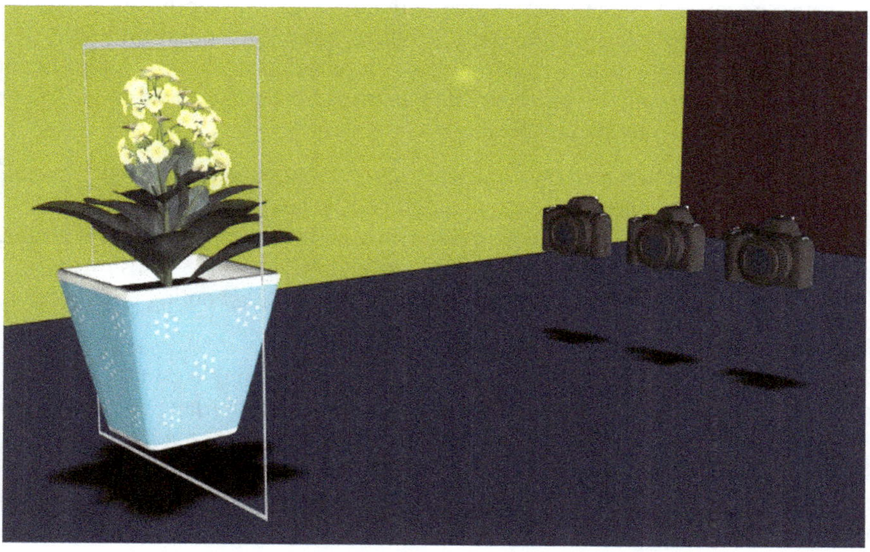

9.1 3D lenticular photography

3D lenticular photography is a process of producing a lenticular sequence by photography. Such process requires shooting from a series of pre-defined points, which can be done with suitable mechanical aids. The resulting sequence of photographs requires some processing before it can be used as a lenticular sequence, which may include projection corrections due to camera misalignment, magnification correction, color/brightness balancing, cropping, etc.

In recent years, specialized algorithms were developed, which allow 3D reconstruction of objects from a series of arbitrary photographs (Hartley & Zisserman, 2003) (Ma, Soatto, Kosecka, & Sastry, 2004). From such reconstruction, it is possible to derive the image sequence using suitable 2D projections. Although this method can be used to derive a lenticular sequence from freely taken photographs, it is not fully reliable.

A 3D lenticular picture which displays the photographed object with correct proportions and size is said to be "ortho-3D". In practice, ortho-3D imaging is rarely used, since the imaged object must comply with the depth design principles. Despite this, it is methodologically useful to start the discussion with rigorous ortho-3D photography, because in this case both the procedure and its outcome are well-defined. Deviations from the ortho-3D procedure and their impact on the outcome will be discussed in the following paragraphs.

There are two primary methods which are used in lenticular photography: parallel and converged. In both methods, each photograph is taken from a different point, called a "shooting point." Rigorously, the shooting points must be located on a straight line with equal distances between them, as illustrated in Figure 113. This rule is followed by the parallel method. The converged method uses shooting points on an arc, which deviates slightly from the straight-line rule.

There are two approaches in 3D lenticular photography: single camera and camera array. With a single camera, there is a need for a mechanism to transport the camera from one shooting point to another. The scene and the illumination need to be perfectly stationary during the whole process, but the translation of the camera from one shooting point to another takes a certain amount of time. The stationarity requirement excludes the photography of dynamic scenes, like running water, with a single camera.

3D lenticular photography with a camera array can be instantaneous, and therefore free from the limitations imposed by the scene stationarity requirement. However, a camera array introduces other difficulties. Two real cameras always differ in their characteristics (like, for instance, the focal length), and these differences must be measured and compensated. The discussion here will be limited to the geometrical aspects of 3D lenticular photography. Such considerations are equally valid for both single and camera array setups.

Nowadays, it is possible to create digital 3D scenes using 3D computer graphics. Like digital images, such scenes do not have any material presence and exist only in the form of digital files. However, all 3D graphics programs possess a camera tool, which can be used to photograph the scene. Both the camera and the photography within the program are virtual. Virtual 3D lenticular photography is much easier compared to real 3D lenticular photography. The virtual camera parameters can be entirely controlled, and, of course, there is no need for any expensive mechanical instrument. The scene itself and its illumination can be held perfectly stationary. The geometrical principles of 3D lenticular photography apply equally to both real and virtual photography.

9.2 The ortho-3D scenarios

9.2.1 The viewing scenario

A viewing scenario for a general viewing distance *V* is shown in Figure 111. The viewing interval is divided into sub-intervals, which correspond to the interlaced image sub-cells. In this example, the number of sub-cells in an interlaced image cell is 10, and the viewing interval length is *S*.

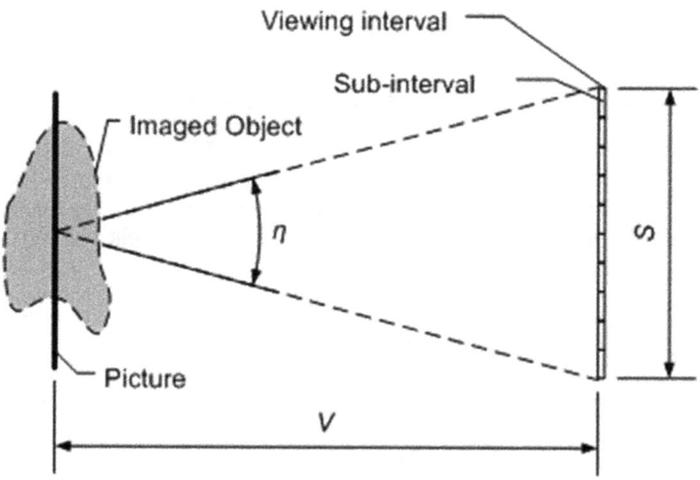

Figure 111: The viewing scenario.

9.2.2 The ortho-3D photography scenario

The ortho-3D lenticular photography scenario is shown in Figure 112. It shows the object and the set of the shooting points which are located on the photography interval. This figure also shows an imaginary object: the picture clone. This is a representation of the picture located in the same position relative to the photographed object as in the viewing scenario (Figure 111). The shooting distance is the same as the viewing distance, and the photography and the viewing intervals lengths are equal.

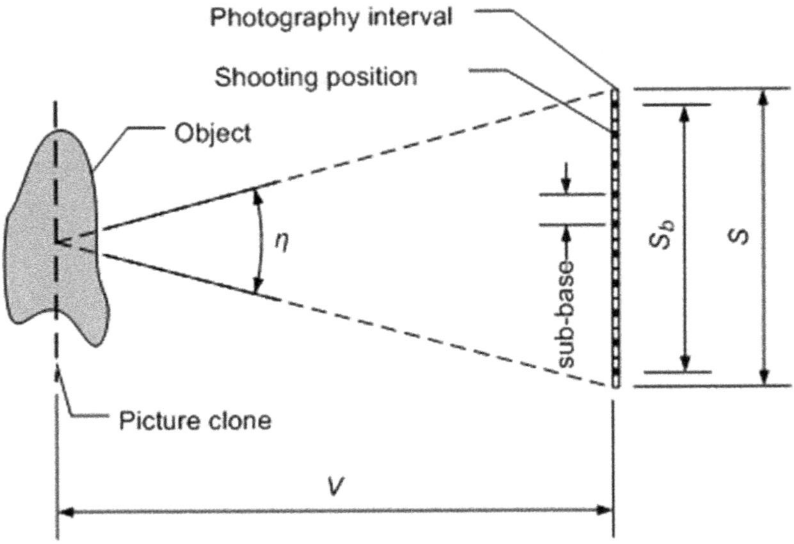

Figure 112: The ortho-3D lenticular photography scenario.

The picture clone object is handy for the explanation and the design of 3D lenticular photography. In fact, it is not needed to clone the whole picture, only its edges. So, the picture clone object is a rectangle outlining the picture edges. In stereoscopy, this object is known as the "stereo window" (Mendiburu, 2009) (Weissman, 2012); here, it will be referred to as "3D window."

In Figure 112, there are 10 shooting points, one for each sub-cell in the main cell of the interlaced picture. The interval between the two extreme points is called the "photography base." In the context of stereoscopic photography, it is called the "stereo base." The distance between adjacent shooting points is called the "sub-base."

The viewing interval length and the photography base are related by

$$S_b = S\left(1 - \frac{1}{N}\right) . \qquad \textit{Equation 99}$$

A 3D lenticular picture made from a sequence of photographs taken from the shooting points shown in Figure 112 will display the object with its real size and proportions, as required by orth-3D display. In practice, 3D lenticular photography is rarely done under these conditions. In most cases an ortho-3D display of the object is not possible, because it will violate the depth design (paragraph 6.7).

9.2.3 Scaled ortho-3D photography scenario

The ortho-3D photography scenario shown in Figure 112 is very limited because quite often the object which one wants to display does not fit in the picture frame. The ortho-3D photography scenario can be scaled to accommodate different size objects

(Weissman, 2012).

Figure 113 (a) shows an ortho-3D photography scenario in which the object is too large to fit the picture. The whole photography scenario is scaled by a factor of 2 (including the picture clone, but without scaling the object) to fit the object in the picture frame. This is shown in Figure 113 (b). After this operation, the imaged object will appear in the picture scaled down by a factor of 2, because the physical picture size remains the same (only the photography scenario is scaled). Although the object is scaled down, its geometrical proportions remain as in the original. This is an example of scaled ortho-3D lenticular photography.

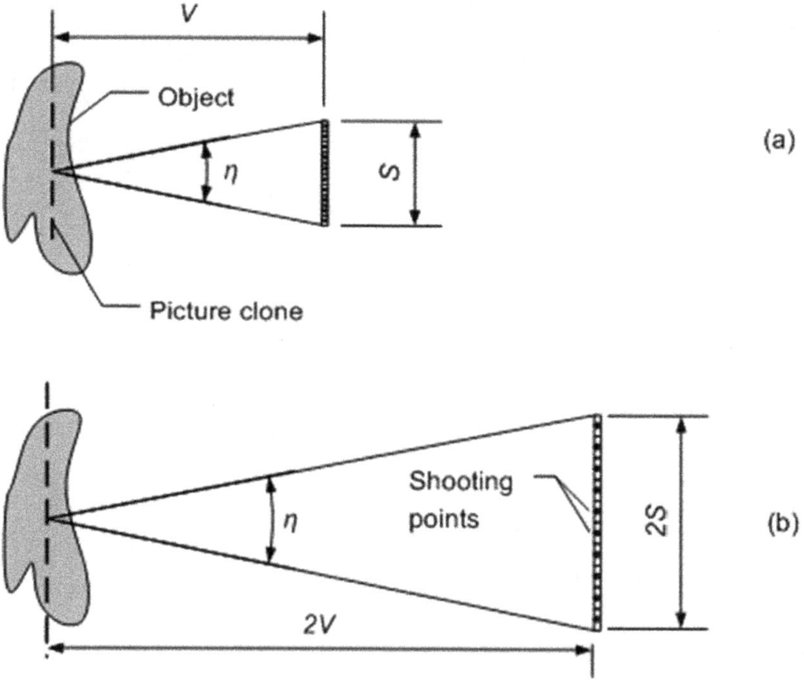

Figure 113: Scaling the ortho-3D photography scenario.

In general, when the photography scenario is magnified (as in Figure 113), the displayed object will be de-magnified. The inverse of the scenario scaling factor can be regarded as the magnification of the object display. In the example shown in Figure 113, the object magnification is 0.5.

Fitting the object to the picture frame by magnification was considered already in paragraph 6.9. Here, it is shown how to control the magnification with 3D lenticular photography.

9.3 Deviations from ortho-3D photography

9.3.1 Shooting distance

In ortho-3D photography, the shooting distance is the same as the viewing distance. In scaled ortho-3D photography, the sooting distance is scaled by the inverse of the object magnification factor. In practice, it is possible to modify the shooting distance in both pure and scaled ortho-3D scenarios with little effect on the imaged geometry. The modification of the shooting distance is illustrated in Figure 114. In this figure, the ortho-3D shooting distance *V* is modified to *V'*. This gives rise to another photography interval with length *S'*, and another set of shooting points. Note that both *S* and *S'* subtend the same angle η with respect to the picture center. This transformation is similar to scaling the photography scenario as discussed in paragraph 9.2.3, but it differs from it by an important aspect: the picture clone is not scaled and remains the same for both *S* and *S'*.

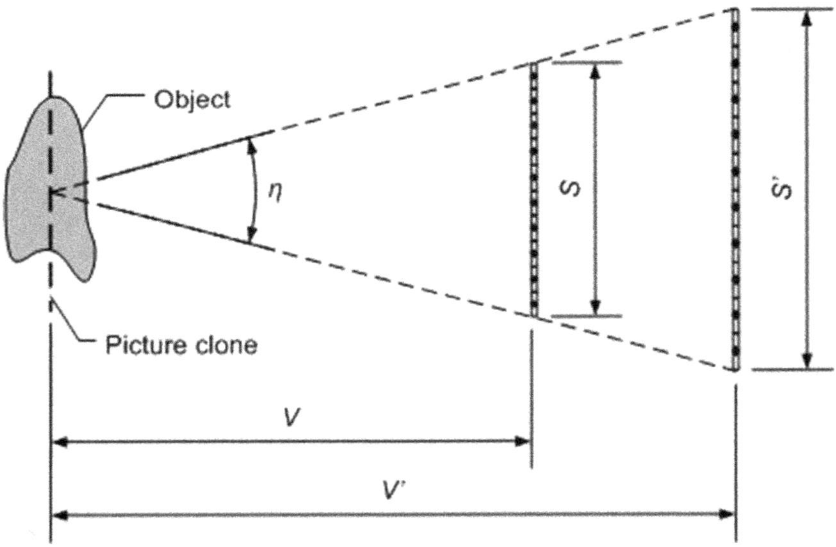

Figure 114: Modification of the shooting distance in 3D lenticular photography.

Changing the shooting distance affects the size of the object in the photograph. This can be offset by adjusting the camera zoom, to make the object size invariant. However, changing the shooting distance causes additional changes in the object appearance, due to the perspective modification, and these changes cannot be offset. However, in most practical conditions, the shooting distance can be significantly modified before perspective induced changes become noticeable. Therefore, shooting at a distance different from the viewing distance but scaling the viewing interval accordingly (as shown in Figure 114) will yield a near ortho-3D picture.

9.3.2 *Controlling the imaged depth*

From Equation 51 and Equation 54 it follows that the distance of an imaged point from the picture plane is proportional to the total parallax *P*. From Figure 50 it can be deduced that *P* is proportional to the photography base used for defining the shooting points. For example, reducing the photography base by a factor of 2 also reduces the imaged depth of the object by the same factor. In converged photography (see paragraph 9.5), depth is controlled similarly with the angular photography base. In this manner, it is possible to fit objects inside the imaging boxes which were introduced in paragraph 6.7 and control the picture visual characteristics. An example of how the imaged depth is controlled with the photography base is given in paragraph 9.8 below.

A general lenticular 3D photography scenario is shown in Figure 115. The photography interval *S'* is positioned at a general distance *V'* from the picture clone plane, and its size is modified compared to the ortho-3D interval to control the displayed depth. Figure 115 also shows the imaged object, whose depth is reduced by the reduction of the shooting interval.

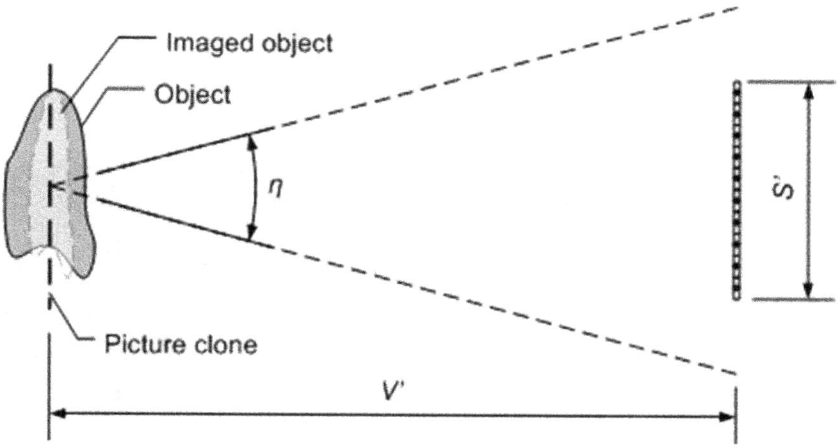

Figure 115: General 3D lenticular photography scenario

9.4 Parallel 3D lenticular photography

In the parallel 3D photography method, the camera must be at the same orientation at all shooting points. This means that it must be displaced parallel to itself when moved from one shooting point to the other. Also, the camera sensor plane must be parallel to the 3D window plane (assuming that it is perpendicular to the camera lens axis). When this condition is satisfied, the 3D window will appear in the photographs as a rectangle. A non-zero angle between these planes will introduce a distortion called a "keystone", and the 3D window will appear in the photograph as a quadrilateral shape. The camera field of view must be large enough to contain the 3D window frame in all shooting positions. The image sequence is derived from the photographs by cropping them to the interior of the 3D window.

There are two main difficulties with the implementation of the parallel method. The first one is due to the parallelism requirement. A high-quality (and expensive) mechanism is necessary to guarantee parallel displacement. The second difficulty originates from the cropping requirement. It is possible to add a real object representing the 3D window to the photographed scene (see paragraph 9.7 below). This object will appear in the photographs, and can be used for precise cropping. Since adding such an object to the scene may be impractical or inconvenient (for instance, because it can cast shadows), it is possible instead to compute its position and perform the cropping based on this computation.

Both difficulties are absent in the implementation of the parallel method in virtual photography. The parallel translation can be carried out with perfect accuracy without any aids. The 3D window can be easily added to the scene and configured so as not to cast shadows or interfere with the scene in any manner. Let us present some illustrations of the parallel method implementation in virtual photography.

Figure 116 shows an artist's view of a virtual parallel photography scenario. Only three shooting positions are shown. The 3D window is represented by a thin white frame. The three photographs taken in the manner described here are shown in Figure 117, and finally, Figure 118 shows the images cropped at the white frame. This procedure can be readily generalized to any number of shooting points, and the resulting sequence may be used for a 3D lenticular picture.

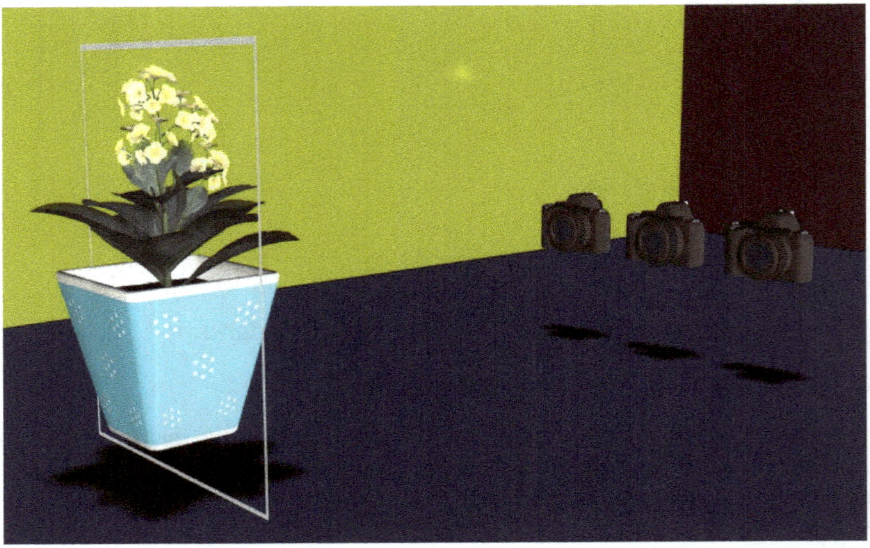

Figure 116: Parallel 3D photography.

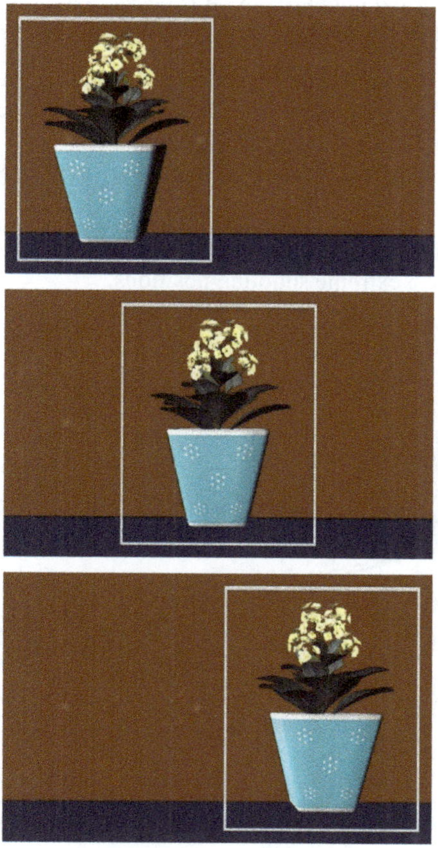

Figure 117: The uncropped images.

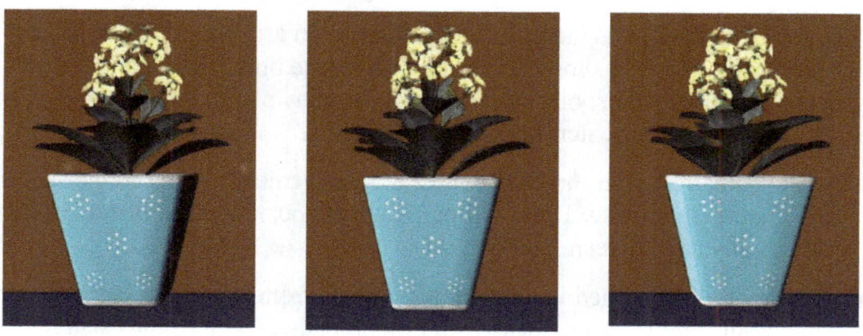

Figure 118: Images cropped to the 3D window frame.

9.5 Converged 3D photography

In the example shown in Figure 117, approximately 50% of the image area is cropped out. In real photography, this reduces the pixel count of the final image, thus wasting precious resources. In virtual photography, it is possible to avoid cropping by using sub-frame rendering, at the expense of complicating the rendering setup.

Cropping waste can be significantly reduced by directing the camera to the 3D window center in each shooting point. The 3D window center serves as a pivot point for the photography. In this case, the sensor plane is no longer parallel to the 3D window plane, thus causing a keystone distortion. However, this distortion can be corrected quite easily.

Converged 3D photography is illustrated in Figure 119. The camera orientation is illustrated only in the two extreme shooting positions.

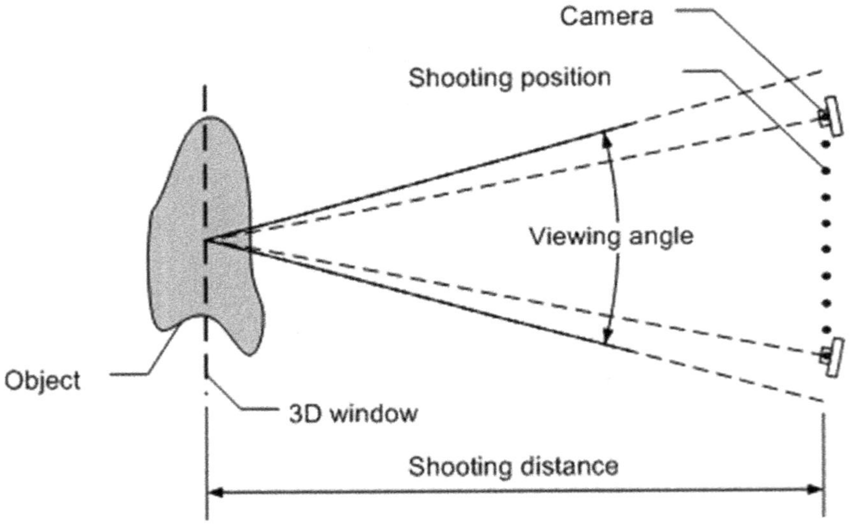

Figure 119: The converged 3D photography method

In the converged method, it is more convenient to use an arc, rather than a straight line to define the positions of the shooting points. This is true both for real and virtual photography. The arc center should be on the 3D window plane center, so the camera is oriented towards the arc center in the shooting points.

When using an arc for the shooting points, it is convenient to refer to the shooting angle as the photography base. The shooting points are positioned so that a constant angle separates any two adjacent points.

The shooting points defined in this manner deviate from a straight line, and they are also not equidistantly spaced linearly. In principle, this fact introduces slight geometric distortions due to change in perspective (as mentioned in paragraph 9.3.1) which cannot be rigorously corrected. However, in all practical cases, these distortions are negligible and can be disregarded.

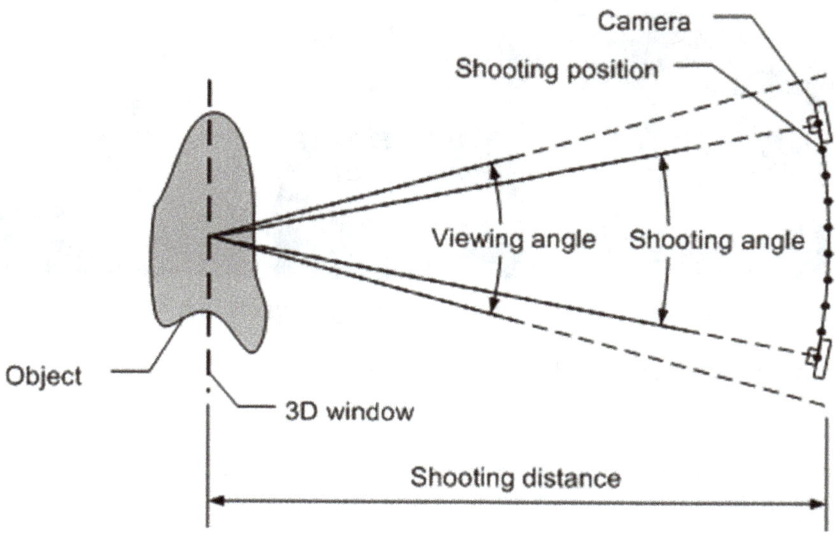

Figure 120: Converged 3D photography using an arc trajectory.

Figure 121 shows the same 3D photography scenario as shown in Figure 116, except that now the cameras are converged to the center of the 3D window. The three resulting photographs are shown in Figure 122. Comparing these pictures to the ones obtained in parallel photography (Figure 117) reveals two differences:

1. The shape of the 3D window is trapezoidal in the two extreme pictures, and not rectangular. This is the manifestation of the keystone distortion.
2. The pixel efficiency is much better.

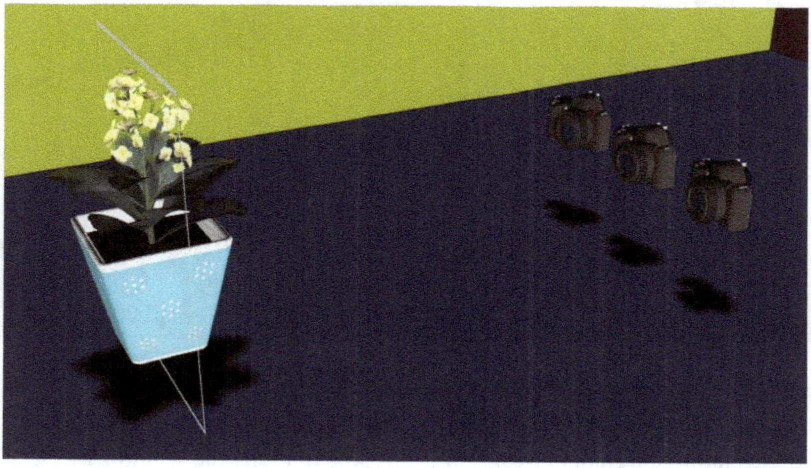

Figure 121: Converged 3D photography.

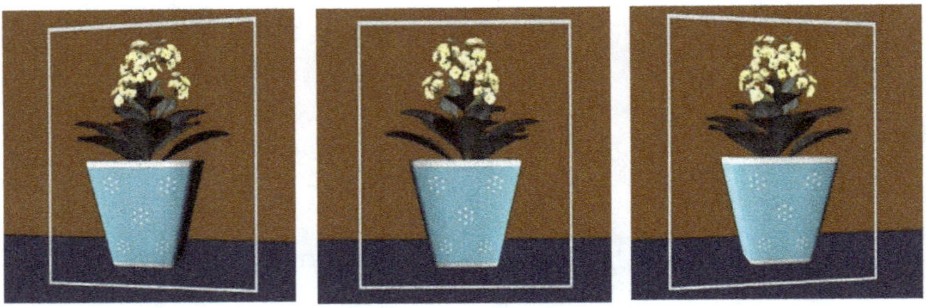

Figure 122: Photographs taken with converged cameras.

9.6 Correction of the keystone distortion

9.6.1 The rotated 3D window

The main tool for the analysis of the keystone distortion is the 3D window object. Figure 123 shows a photography scenario with two cameras and the 3D window object only. The first camera is centered, namely its optical axis passes through the window center, which serves as the pivot point. The other camera is created by rotation of the centered camera in the horizontal plane by 45° around the pivot point. In real applications rotation angles are typically less than 6°, but here an abnormally high rotation angle is used to increase and illustrate the distortion.

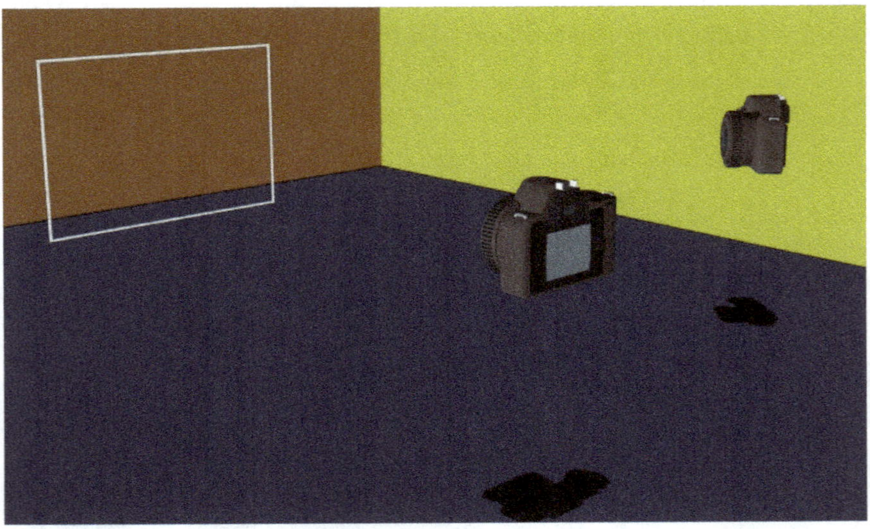

Figure 123: Converged photography scenario for the illustration of the keystone distortion.

Instead of rotating the camera by 45°, it is possible to rotate the window by -45°. This scenario is shown in Figure 124. The straight and the rotated windows are shown superimposed; in reality, there is only one window.

Photography using a rotating object and a stationary camera is often used as an alternative to a stationary object and a rotating camera. Photographically, the results of both methods are identical. The only difference is that the rotating camera scenario can be implemented with a camera array and shot simultaneously from all shooting positions. Simultaneous shooting is impossible with a rotating window; each shot must be taken separately.

The analysis of the keystone distortion is easier for the rotating window scenario, and will be used from now on. The actual photograph corresponding to the scenario of Figure 124 is shown in Figure 125. The photograph plane in Figure 125 is the straight window plane.

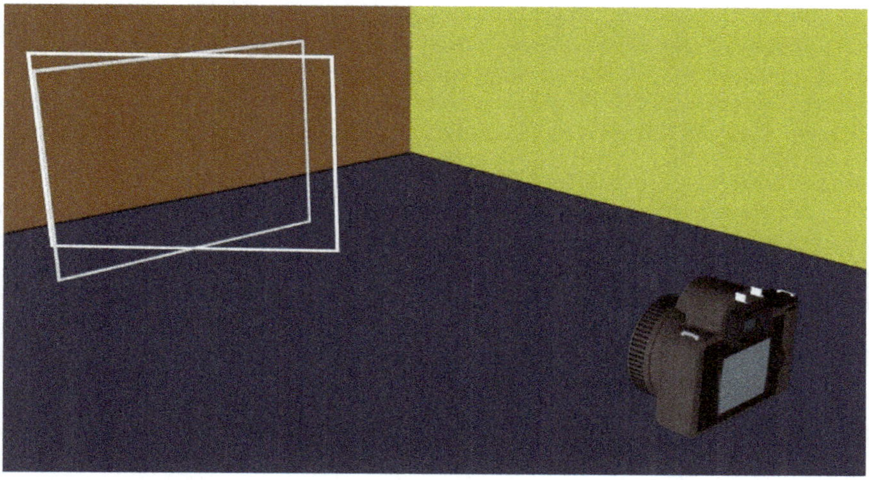

Figure 124: Converged photography scenario with a stationary camera and rotated window.

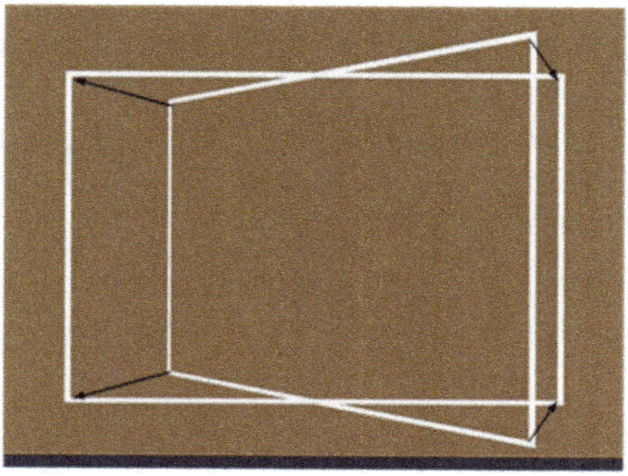

Figure 125: Superimposed photograph of the straight and the rotated windows.

Keystone distortion is corrected by a projective transformation. Such a transformation is uniquely defined by two sets of coordinates: reference and distorted (paragraph 3.9). The reference coordinates define the straight window corners, and the distorted coordinates define the rotated window corners, so each set contains coordinates of four points. All coordinates are measured in the plane of the photograph, which is the straight window plane. The projective transformation will map the corners of the distorted (rotated) window on the corners of the reference (straight) window, as shown by the arrows in Figure 125. After this transformation, the rotated window image will become identical to the straight window image, and the keystone distortion will be corrected.

Unfortunately, common image processing programs do not offer a general projective transformation tool. Such tools are included in more technically oriented software packages, like the image processing libraries of Matlab (Mathworks, 2017) and OpenCV (OpenCV, 2017). If the coordinate sets of the projective transformation are known, they can be executed automatically with a suitable computer program.

The application of such automatic keystone correction is possible only if the camera orientation is controlled with high accuracy. In real photography, such control can be achieved with a high-accuracy mechanical device. Virtual photography is always perfectly controlled, and therefore automatic keystone correction is particularly suitable for computer-generated scenes.

9.6.2 Computation of the rotated window corners coordinates

The computation of the corner coordinates of the rotated window involves only straightforward trigonometry but can be confusing because it is done in three dimensions. To help the reader, the exposition is divided into two stages. In the first stage, the coordinates are calculated with real dimensions on a photograph printed with a unity magnification. This is helpful because thinking in terms of real objects is easier than considering abstract ones. In practice, the projective transformation is done on the photographed image. Therefore, pixel coordinates are needed rather than real coordinates. The conversion from real coordinates to pixel coordinates is done in the second stage.

The diagram used to compute the corners coordinates is shown in Figure 126. This diagram is a top view of the photography scenario. The point ***O*** is the camera projection center, L is the shooting distance, and α is the rotation angle.

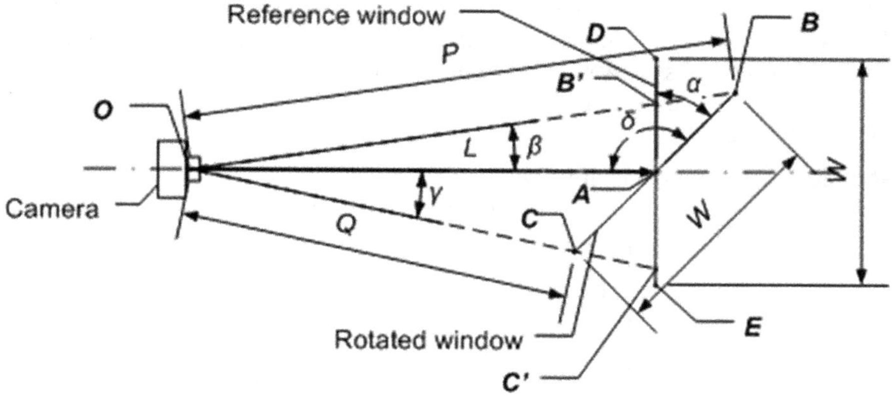

Figure 126: Diagram for the computation of the corner coordinates in the rotated window.

Applying the cosine theorem for the triangle OAB yields

$$P^2 = L^2 + W^2 - LW\cos\delta\ .$$

Since $\delta = 90° + \alpha$, this can be rewritten as

$$P^2 = L^2 + W^2 + LW\sin\alpha\ . \qquad \text{Equation 100}$$

Using the sine theorem for the same triangle gives the relation

$$\sin\beta = \frac{W}{2P}\cos\alpha\ . \qquad \text{Equation 101}$$

Combination of this with Equation 100 yields

$$\sin\beta = \frac{\cos\alpha}{2\sqrt{1+\left(\frac{L}{W}\right)^2+\frac{L}{W}\sin\alpha}}\ . \qquad \text{Equation 102}$$

Similar reasoning leads to the analogous relation for γ:

$$\sin\gamma = \frac{\cos\alpha}{2\sqrt{1+\left(\frac{L}{W}\right)^2-\frac{L}{W}\sin\alpha}}\ . \qquad \text{Equation 103}$$

Figure 127 presents the photograph shown in Figure 125 with the addition of a few annotations. The points ***A*, *B'*, *D*, *C'*, *E*** appear both in Figure 126 and in Figure 127. Point ***A*** serves both as the pivot point for the photography, and as the origin for the photograph system of coordinates, as shown in Figure 127.

Let us assume that the photograph of the straight and the rotated windows was printed with unity magnification, so that the straight window appears in the printed photograph as its real size. A reduced picture of this print is shown in Figure 127. The x-coordinates of the points in Figure 127 are shown in Table 23. The y-coordinates of all these points is 0.

Table 23: Coordinates of the points of Figure 127

Point	x-coordinate
A	0
B'	$-L\tan\beta$
D	$-W/2$
C'	$L\tan\gamma$
E	$W/2$

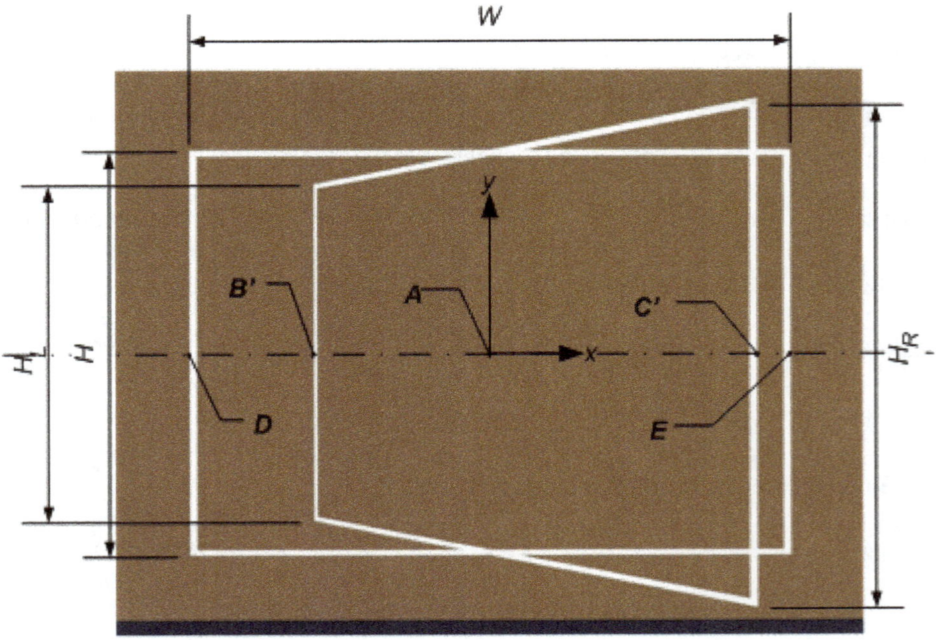

Figure 127: The windows photograph annotated.

The rotated window appears as a trapezoid with left edge length H_L, and right edge length H_R. These are the projections of the left and the right edges respectively of the rotated window on the straight window plane.

To compute H_L, let us consider the plane defined by the left edge of the rotated window and the camera projection point ***O***. This plane, shown in Figure 128, contains the points ***B*** and ***B'*** which also appear in Figure 126 and Figure 127.

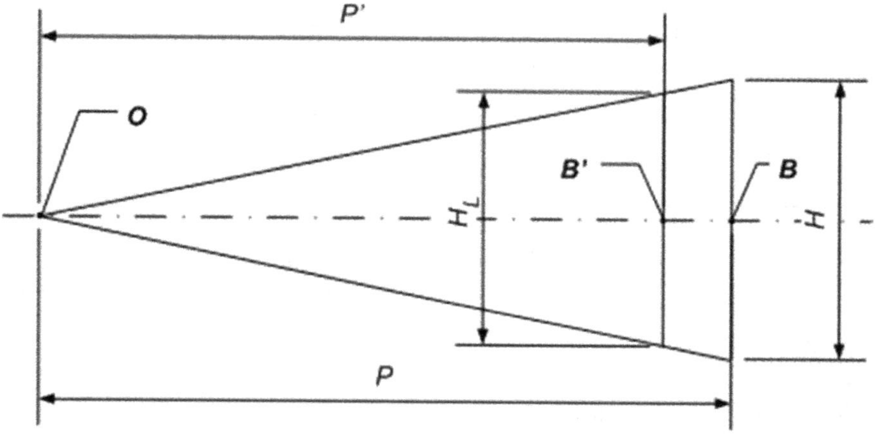

Figure 128: Diagram for the computation of H_L.

From Figure 126 it follows that

$$P' = \frac{L}{\cos(\beta)}.$$

Using triangular similarity

$$\frac{H_L}{H} = \frac{P'}{P}.$$

Using Equation 101, this can be rewritten as

$$\frac{H_L}{H} = \frac{2L}{W}\frac{\tan\beta}{\cos\alpha}. \qquad \textit{Equation 104}$$

Similarly, the following relation for H_R is obtained:

$$\frac{H_R}{H} = \frac{2L}{W}\frac{\tan\gamma}{\cos\alpha}. \qquad \textit{Equation 105}$$

The coordinates of the rotated and the straight windows are shown in Table 24. Note that for a given scenario (given L, W, and H), the transformation still depends on the rotation angle α. Thus, the coordinates must be computed for each rotation angle value, and there will be a different transformation for each such angle.

Table 24: Corners coordinates in real dimensions.

Corner	Straight window		Rotated window	
	x	y	x	y
Top left	$-W/2$	$H/2$	$-L\tan\beta$	$\frac{LH}{W}\frac{\tan\beta}{\cos\alpha}$
Bottom left	$-W/2$	$-H/2$	$-L\tan\beta$	$-\frac{LH}{W}\frac{\tan\beta}{\cos\alpha}$
Bottom right	$W/2$	$-H/2$	$L\tan\gamma$	$-\frac{LH}{W}\frac{\tan\gamma}{\cos\alpha}$
Top right	$W/2$	$H/2$	$L\tan\gamma$	$\frac{LH}{W}\frac{\tan\gamma}{\cos\alpha}$

It is instructive to see what happens to the keystone distortion when $L \to \infty$. From Equation 102 and Equation 103 it can be deduced that in this limit

$$L \tan\beta = L \tan\gamma = \frac{W \cos\alpha}{2} \; .$$

Substituting this in Table 24, the coordinates of the rotated window for $L \to \infty$ are derived. The results are shown in Table 25. It is seen that in this limit the vertical displacements of the rotated window corners vanish, but the horizontal displacements do not.

Table 25: Corners coordinates in real dimensions for an infinite shooting distance.

Corner	Straight window		Rotated window	
	x	y	x	y
Top left	$-W/2$	$H/2$	$-\frac{W\cos\alpha}{2}$	$H/2$
Bottom left	$-W/2$	$-H/2$	$-\frac{W\cos\alpha}{2}$	$-H/2$
Bottom right	$W/2$	$-H/2$	$\frac{W\cos\alpha}{2}$	$-H/2$
Top right	$W/2$	$H/2$	$\frac{W\cos\alpha}{2}$	$H/2$

The real coordinates need to be converted to image pixel coordinates. To accomplish this, it is necessary to know the camera pixel size when projected on the straight window plane. This can be determined from the photograph by measuring the pixel size of the straight window width. Denoting this pixel size by W_p, we get the following value for the projected pixel size q:

$$q = \frac{W}{W_p} .$$

Any physical dimension which was derived, like the coordinates in Table 24, can be converted to pixel dimensions by dividing it by q. The converted corner coordinates of Table 24 are given in Table 26. For convenience, the ratio L/q was denoted by ϕ.

Table 26: Corners coordinates in pixel dimensions

Corner	Straight window		Rotated window	
	x	y	x	y
Top left	$-\frac{W\phi}{2L}$	$\frac{H\phi}{2L}$	$-\phi\tan\beta$	$\frac{H\phi}{W}\frac{\tan\beta}{\cos\alpha}$
Bottom left	$-\frac{W\phi}{2L}$	$-\frac{H\phi}{2L}$	$-\phi\tan\beta$	$-\frac{H\phi}{W}\frac{\tan\beta}{\cos\alpha}$
Bottom right	$\frac{W\phi}{2L}$	$-\frac{H\phi}{2L}$	$\phi\tan\gamma$	$-\frac{H\phi}{W}\frac{\tan\gamma}{\cos\alpha}$
Top right	$\frac{W\phi}{2L}$	$\frac{H\phi}{2L}$	$\phi\tan\gamma$	$\frac{H\phi}{W}\frac{\tan\gamma}{\cos\alpha}$

9.6.3 *Application of the automatic correction: an example*

The photographs shown in Figure 122 were made by a virtual camera in a computer-generated 3D scene. The orientations at the three shooting positions were perfectly controlled, and the convergence angle is precisely known (11° in this case). Accordingly, the keystone distortion was corrected by a computer, and the result is shown in Figure 129. Next, the photographs were cropped to the 3D window interior, to derive the final form shown in Figure 130. These photographs are nearly identical to the ones obtained with the parallel photography method (Figure 118).

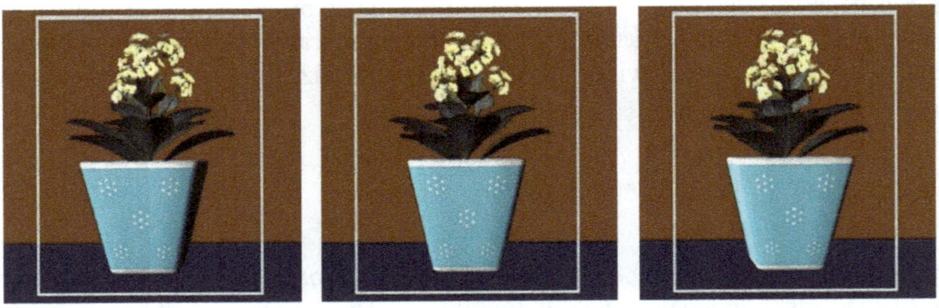

Figure 129: Converged photographs with keystone distortion corrected.

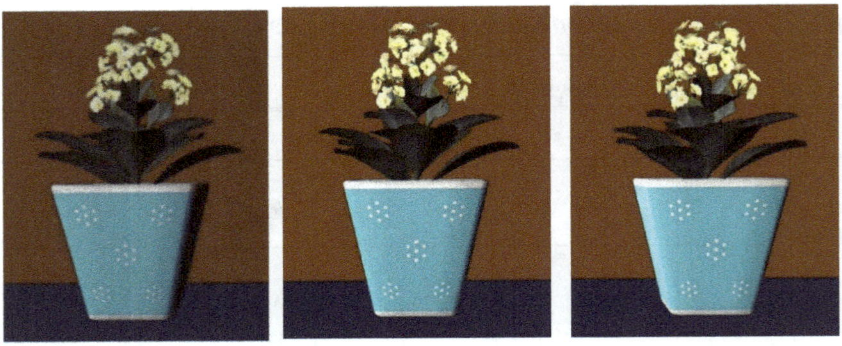

Figure 130: Converged photographs in final form (with keystone correction and cropping).

9.7 Real photography example

9.7.1 Introduction

This paragraph shows an example of processing three photographs made with a real camera array. The camera rig was not a high-accuracy instrument, and the camera orientations deviate from the ideal values by unknown angles. The only precisely controlled parameters were the cameras locations. Also, there are small variations in the camera's magnifications, due to small variations in their focal lengths. All of these lead to more general distortions than a simple keystone, which nevertheless can be corrected by suitable projective transformations. This correction is called "registration." In the present example, the registration will be assisted by including a rectangular frame in the scene, which represents the 3D window.

There are readily available computer programs capable of performing automatic registration, even without the presence of a 3D window in the photographs. Perhaps the most popular programs offering this option are Stereo Photo Maker (Stereo Photo Maker, 2017) and Cosima (Cosima, 2017). Such automatic registration is based on specific algorithms which work well in most of the cases but fail in some. Thus high-accuracy registration cannot be guaranteed in automatic programs. The only way to guarantee registration in all cases is to rely on a real 3D window in the scene.

Registration with the 3D window in the scene can be assisted by a suitable computer program, in which the frames corner coordinates can be measured, to carry out the required transformation. Such a program is not commercially available at the time of the writing of this book. Here it will be shown how to correct the distortions manually using a familiar image editing software.

The raw photographs of this example are shown in Figure 131[4]. They show a model resting against a white background wall, on which a rectangular frame is mounted, to serve as a physical representation of the 3D window. The four corner marks on the frame denote the corners of the 3D window. The manual process consists of three steps:

1. Cropping the raw photographs to the 3D window frame,
2. Registration of the cropped raw photographs,
3. Cropping the registered photographs to the frame interior.

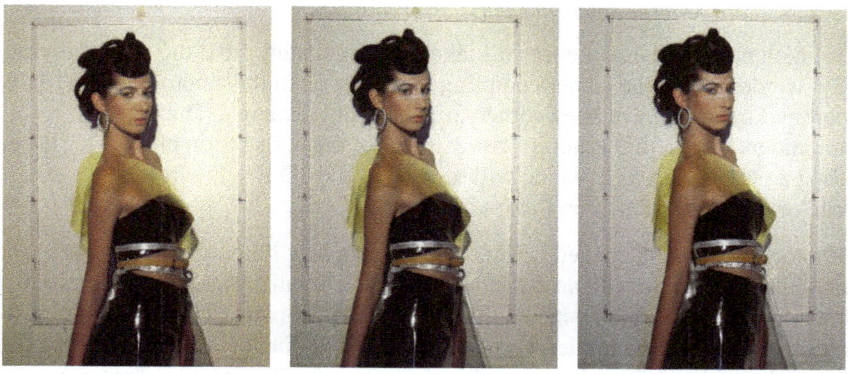

1 2 3

Figure 131: The raw photographs.

9.7.2 Initial cropping

The first step consists of cropping the raw photographs in Figure 131 to the exterior of the frame. The outcome of this cropping is shown in Figure 132. This is necessary because manual registration described here works best when the image corners are close to the window corners.

[4] Courtesy of Oren Or Kobi Hair Design

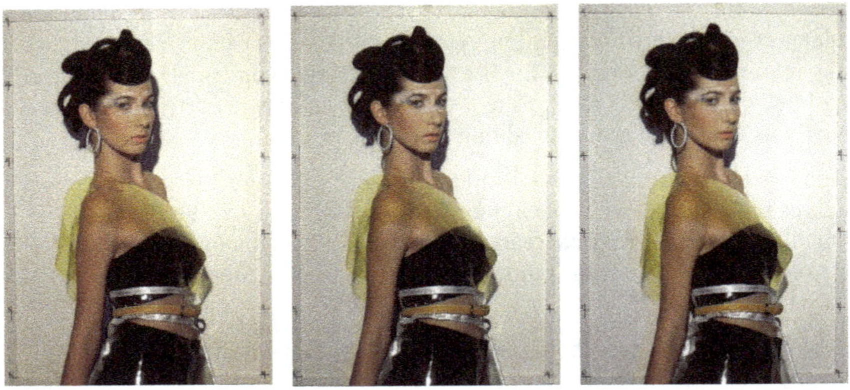

1 2 3

Figure 132: The raw photographs cropped.

9.7.3 Creation of the reference window rectangle

Photograph no. 2 in the sequence was taken approximately from the central normal of the 3D window, with the camera pointing to its center. In this photograph, the four corner marks should have been corners of an upright rectangle. Due to small misalignment errors, this is not precisely true. So, the registration process is started by creating a perfect rectangle which will be used as the 3D window for the registration instead of the corner marks on the frame.

Photograph no. 2 is opened in Photoshop, and a new transparent layer is added. Then a black rectangle with an aspect ratio of the lenticular picture (in this case 0.71) is drawn on this layer and resized until it best fits the corner marks on the frame. Then it is translated so that its top left corner coincides with the top left corner of the frame in the photograph. The result is shown in the left part of Figure 133. This rectangle will serve as the reference 3D window for the registration process.

The top right and bottom left corners fit almost perfectly, and the only significant deviation is in the bottom right corner. A magnified view of the corners is shown in Figure 134.

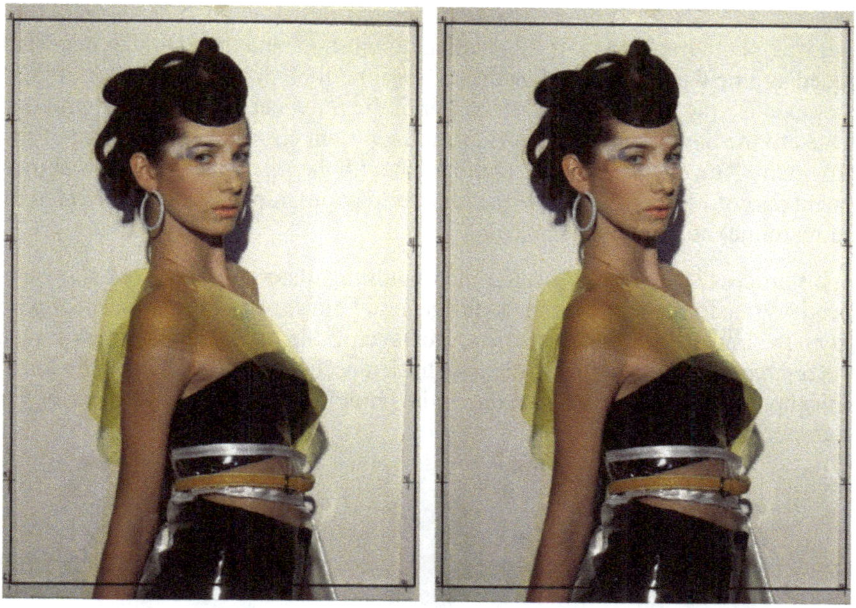

Original before registration Adjusted after registration

Figure 133: Photograph registration with the reference window.

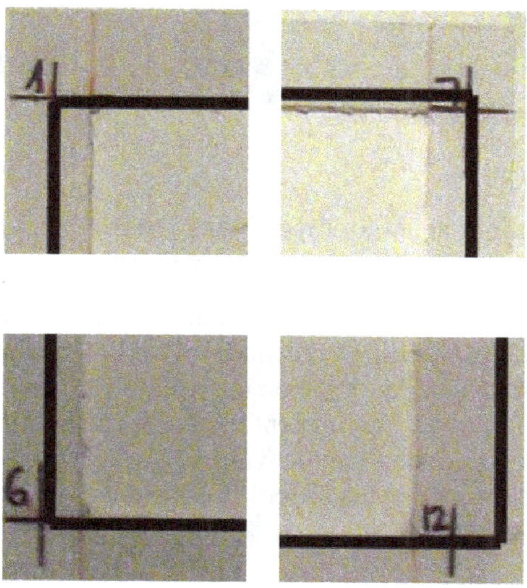

Figure 134: Corners misalignment of photograph no. 2 (before registration).

9.7.4 Registration

To register photograph no. 1, it is opened in Photoshop, and the 3D reference window is added as a new layer. Using the editing "Distort" tool, the top left corner of the photograph is dragged to the top left corner of the reference window. This process is repeated to the bottom left, bottom right, and top right corners, and to the other two photographs. The adjusted second photograph with the reference window is shown in the right part of Figure 133. The three registered photographs (without the reference window frame) are shown in Figure 135.

This process is not exact, because the Photoshop distort tool does not allow a projective transformation to be made for general points; one of the two sets of points must be the transformed image corners. However, if the photograph corners are close to the reference window corners, the registration accuracy will be acceptable for all practical purposes. For this reason, the initial cropping is performed as described in paragraph 9.7.2.

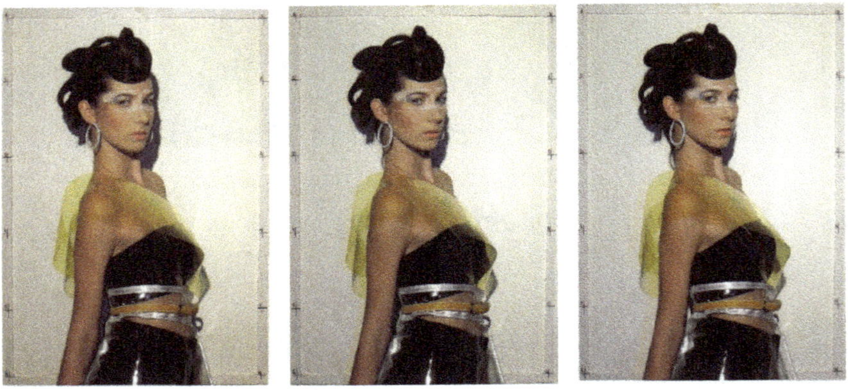

Figure 135: Registered photographs.

9.7.5 Final cropping

The registered photographs are now cropped to the frame interior, so that it does not show in the lenticular picture. The final photographs are shown in Figure 136.

Note that in the final sequence shown in Figure 136, the 3D window plane is coincident with the wall plane behind the model. This is done for practical reasons, as it is most convenient to mount the frame on the wall. As explained in paragraph 6.7, such disposition does not allow the full depth imaging capabilities of the lenticular picture to be used. The position of the 3D window plane can be easily adjusted after the registration process is completed, as will be shown below.

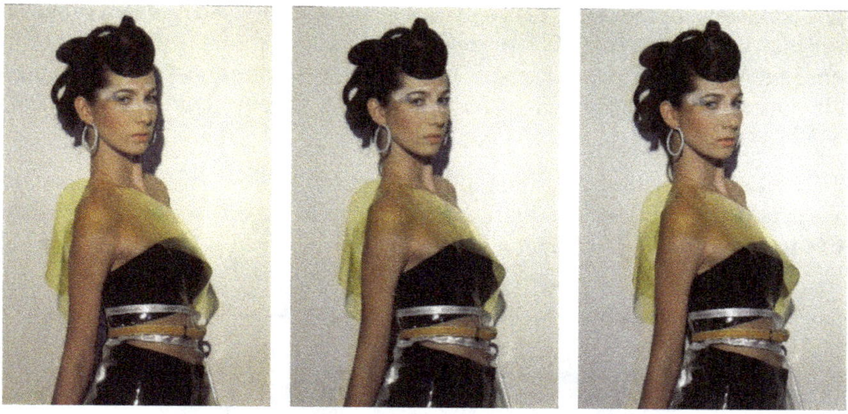

Figure 136: Cropped and registered (final) photographs.

9.8 Photography and depth design

The depth design (Chapter 6) has two outcomes which determine the photography scenario: the number of images in the sequence and the parallax factor x (Equation 63). The parallax factor depends on the imaged object point. In the present discussion, the parallax factor will correspond to the object point which is furthest from the picture plane. This parallax factor determines the maximal incremental parallax in the picture, $P_{s,max}$:

$$P_{s,max} = xp \, ,$$

where p is the lenticular pitch.

The incremental sequence parallax P_s is proportional to the shooting interval length S' or the shooting angle. This fact can be used to adjust the incremental parallax to the desired value.

Let us imagine that the image sequence was printed to obtain a sequence of pictures with the same sizes as the planned lenticular picture. In such a sequence, one can measure the incremental parallax P_s for any object point in physical units. It is convenient to introduce the relative parallax δ:

$$\delta = \frac{P_s}{W} \, ,$$

where W is the picture width. The largest relative parallax is given by

$$\delta_{max} = \frac{P_{s,max}}{W} = \frac{x}{N_L} \, ,$$

where N_L is the number of lenticules in the picture.

The advantage of the relative parallax concept is that it is independent of the actual picture dimension. Therefore, it can be measured directly from the pixel sizes of the incremental parallaxes in the sequence images. Let us show how to determine the angular increment for converged 3D photography for the object used in paragraphs 9.4 and 9.5.

Figure 137 shows a superposition of the two extreme views of Figure 130. The largest parallax size in pixels is 30, and the image width in pixels is 562. This gives a relative parallax value of 0.0534.

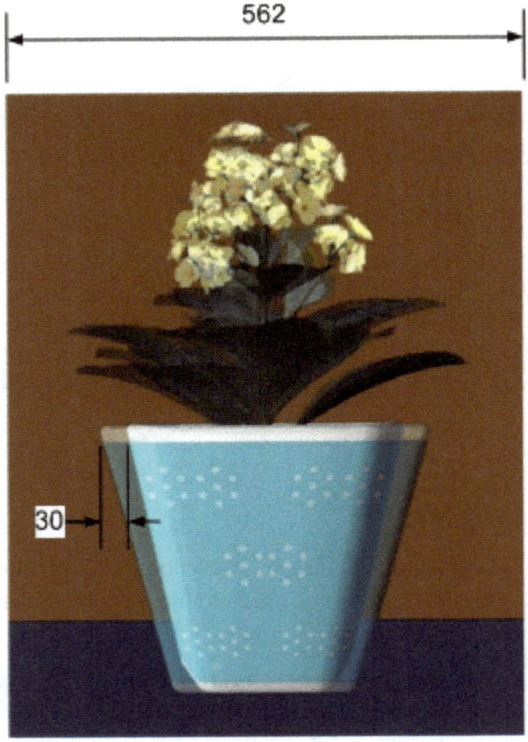

Figure 137: Measurement of the relative parallax.

Suppose that one wants to make a 10" wide picture of this object using a 60lpi lenticular sheet. The number of lenticules in such a picture will be 600. Assuming that $x = 2$ was chosen, the maximal relative incremental parallax must be less than 0.00333, and therefore the angle increment between such images must be reduced by a factor of 0.06 compared to the angular separation between the two images in Figure 130. Since the angular separation in Figure 130 was 22°, the angular separation between adjacent images should be 1.37°.

9.9 Adjustment of the 3D window plane position

The parallax of points in the 3D window plane is zero. Therefore, these points appear sharpest in the picture. If there is an interesting texture with subtle details in the scene, it is preferable to put it on the 3D window plane. If the picture is a portrait, one may wish to put the subject's face on the 3D window plane.

The choice of the 3D window position may have a significant impact on the picture visual quality. The auxiliary frame which was used in the photography scenarios discussed in this section determines the position of this window. Thus, for example, in the real photography example discussed in 9.7, the 3D window plane will coincide with the background wall.

In virtual photography, it is possible to position the auxiliary frame anywhere, because it does not need any physical support, and can be configured not to cast shadows. In real photography, there are practical constraints which may force the photographer to mount the auxiliary frame in a non-optimal position. However, the 3D window plane can be adjusted by simple post-processing. The method is illustrated in Figure 138.

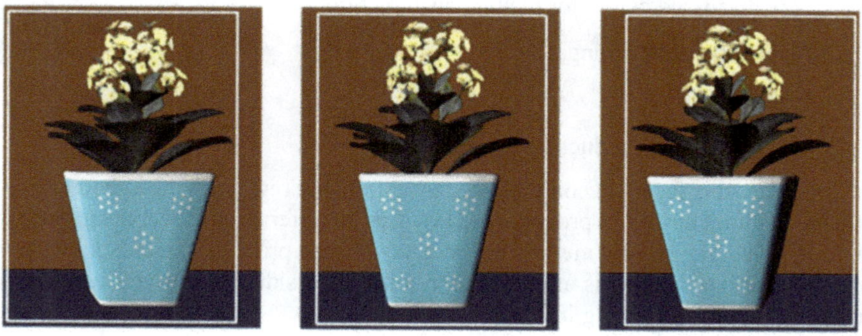

Figure 138: Adjustment of the 3D window plane using a synthetic window in post-processing.

One can always add objects to the sequence photographs using simple graphical techniques. It is possible to compute the position of an imaginary frame disposed at any given position in the scene for each photograph and artificially superimpose it on the sequence photographs. Figure 138 shows a sequence processed in this manner, with an imaginary window behind the real window of the sequence. Since the imaginary window is not on the 3D window plane, it will show a parallax. Cropping the photographs shown in Figure 138 to the interior of the imaginary window will create a new sequence with the position of the 3D window translated backward to the plane of the imaginary window.

9.10 Increasing the continuous depth display

9.10.1 General

The general principles of depth design were discussed in paragraph 6.11, where a depth factor z was introduced. This factor is the ratio between the imaged depth and the resolved imaging box depth. It is common to design lenticular pictures with $z > 1$, in order to display more depth. Increasing depth beyond the resolved imaging box introduces either blur or discretization in the picture.

As discussed in paragraph 6.11, the depth design is determined by two parameters, which were called x and y. The product of these parameters equals the depth enhancement factor (Equation 66). A design with $x = z$ and $y = 1$ gives a sharp (without blur) picture but with discretization. In this design, the sequence length is N_r, as listed in the second entry of Table 18. The printing of a picture corresponding to such sequence does not require further elaboration.

In the fully continuous design (the first entry in Table 18), the sequence length N exceeds N_r. The printing of a picture with $N > N_r$ requires special consideration. Three methods are discussed below for making such picture:

1. Over-resolved printing,
2. Blur emulation,
3. Sequence length reduction by image processing.

The first method is the most common, because it fits the regular picture production workflow without any extra processing. However, it suffers from several shortcomings and limitations. The second method relieves most of the problems associated with the first. However, both methods introduce extra blurring inside the resolved imaging box near the edges of the viewing interval.

The third method resolves all problems associated with the first two methods and retains the sharp display within the resolved imaging box over the whole viewing interval.

9.10.2 Over-resolved printing

In digital print, the limiting factor in the picture resolution is the printer resolution (paragraph 6.6, Chapter 10). Therefore, in digital print, sequences with $N > N_r$ generate interlaced images which need to be printed at densities exceeding the printer's resolution. Such printing will be called "over-resolved printing." Over-resolved printing causes the graphical content of y adjacent sub-cells of the interlaced image to blend into each other, thus creating a blur in the lenticular picture. As explained in paragraph 6.7, blurring of object points beyond the resolved imaging box is inevitable. Therefore, this effect is not considered as a shortcoming of over-resolved printing. Nevertheless, printing of an interlaced image with $N > N_r$ has two significant disadvantages:

1. There is a limit to the magnitude of printing density which can be specified. This limit sets a bound on N, and therefore on the value of continuous imaging depth.

2. The interlaced image file size increases linearly with N. This may cause difficulties if the file size increases beyond a certain limit.

9.10.3 Blur emulation

Both problems can be resolved by emulating the printer blurring effect by suitable image processing before print. Such processing can be applied either to the continuous or the rasterized interlaced image. Both approaches lead to similar results, but, this way or the other, the final interlaced image will be blurred by two independent mechanisms: rasterization and print blur emulation.

The details of print blur emulation processing will not be discussed here; qualitatively, this is a transformation which blends together y adjacent sub-cells and merges them into a single sub-cell. The horizontal pixel count of the processed interlaced image will be reduced by such operation from $\{N_L Nc\}$ to $\{N_L N_r c\}$ pixels. The reduced interlaced image can be printed now at a resolved printing density, but the outcome will be like over-resolved printing of the original (unprocessed) image.

In this method, the horizontal pixel count of the processed interlaced image is $\{N_L N_r c\}$, which is independent of N. The vertical pixel count is determined by the picture height and the vertical pixel density, as explained in paragraph 4.5, and is also independent of N. Therefore, in theory, there is no limit on the value of N, and, consequently, there is also no limit on the continuous imaging depth that can be displayed. Furthermore, the interlaced image file size is also reduced and becomes independent of N.

9.10.4 Intracell and intercell blurring

Both over-resolved and emulated printing introduce a certain amount of blurring inside the resolved imaging box because they exhibit both intracell and intercell blurring. These concepts are illustrated in Figure 139 for the case $y \sim 4$.

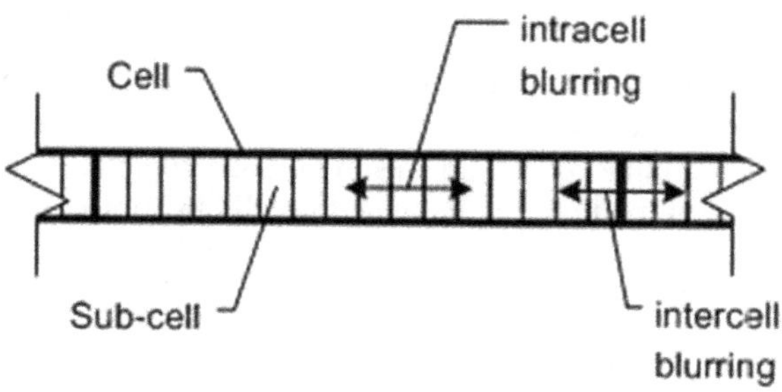

Figure 139: Intracell and intercell blurring.

Intracell blurring mixes the graphical content of y adjacent sub-cells from the same cell. Such sub-cells represent sequential images. The maximal incremental parallax within the resolved imaging box of images with $x = 1$ is p/y, and therefore, inside this box, intracell mixing will increase their incremental parallax to p. As explained in paragraph 6.7, object points with incremental parallax which is less than p appear sharp. Therefore, intracell blur does not affect the display inside the resolved imaging box.

Intercell blurring mixes the graphical content of sub-cells belonging to different cells, which correspond to non-sequential images. The maximal incremental parallax of y non-sequential images is generally much larger than p. Therefore, intercell blurring increases the incremental parallax beyond p and causes blurring inside the resolved imaging box. This blurring will degrade the visual quality of the picture.

Intercell blurring cannot be avoided in both over-resolved and emulated printing methods discussed above.

9.10.5 Sequence processing

Sequence processing resolves all the problems associated with the over-resolved and emulated printing methods. In particular, it completely avoids intercell blurring. It also allows a preview of the images displayed by the picture before it is printed. Due to these advantages, sequence processing is the preferred method for increasing the limit of continuous imaging.

The sequence processing algorithm uses the original sequence of length N to create a new sequence of length N_r, which will be called "reduced sequence." The original and the reduced sequences will be denoted by $\boldsymbol{G}$ and $\boldsymbol{H}$ respectively. The sequence $\boldsymbol{H}$ can be interlaced and printed at a resolved printing density so that there will be no printer blur. In particular, the damaging intercell blurring will be avoided entirely. The lenticular picture will display the images of the sequence $\boldsymbol{H}$, which can be inspected before the actual printing.

A sequence of length N_r can be created by sampling the sequence $\boldsymbol{G}$ at intervals of y images. The number of images in such sequence will be N/y, which is N_r, as required. Its maximal incremental parallax will be y times the maximal incremental parallax of $\boldsymbol{G}$, and it will correspond to the fully sharp design (second entry in Table 18).

A sequence which can be used for a fully continuous design can be created by regarding each image of $\boldsymbol{H}$ as representing y images of $\boldsymbol{G}$. Thus, each image of $\boldsymbol{H}$ should be a certain combination of y images of $\boldsymbol{G}$. Such combination will introduce a horizontal blur, like a motion blur, in the images of $\boldsymbol{H}$, and object points will become elongated spots. These heuristic arguments can be formulated in a mathematical algorithm, which will be described with the help of the diagram shown in Figure 140.

This figure shows the shooting interval and points of the original sequence. For the sake of the algorithm, the set of the shooting points must be extended beyond the shooting interval, as shown in the figure. The whole line of the shooting interval is covered with an infinite number of shooting points, all equally spaced. The original sequence is extended in this manner to an infinite sequence, corresponding to the infinite set of the shooting points. This infinite sequence will be denoted by ***G'***. Furthermore, a linear coordinate u is defined on the shooting interval line. The origin of this coordinate is the center of the shooting interval.

Other objects shown in Figure 140 are the original sequence sub-interval, and a set of points called "reduced sequence points." These points lie at the centers of larger sub-intervals denoted by thicker lines. There are exactly N_r reduced sequence points, and there will be a unique reduced sequence image associated with each point. In the example shown in Figure 140, $N = 20$ and $N_r = 5$, so $y = 4$.

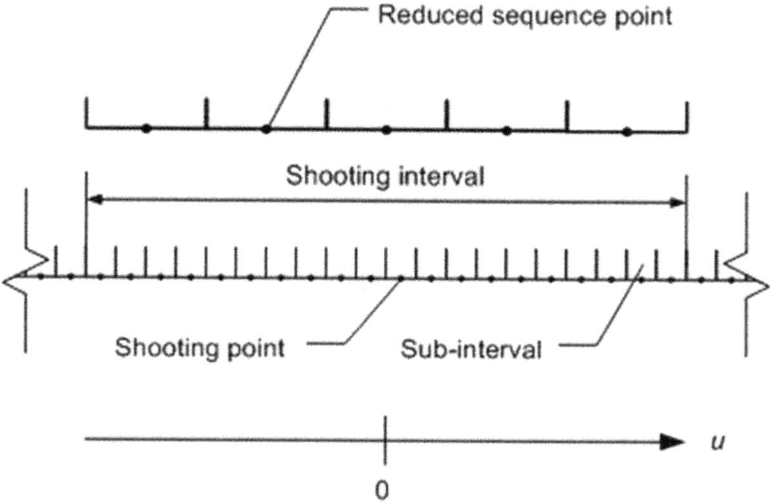

Figure 140: Processing image sequence for extended continuous imaging.

Let us denote the set of shooting points coordinates by s_k, where the range of k is $(-\infty, +\infty)$, and the set of the processed sequence points coordinates by v_l, where the range of l is $[1, N_r]$. For the mathematical formulation, it is convenient to regard the images of the sequences as being a continuous function of the u coordinate. With these conventions, the processed sequence ***H*** is defined by the formula

$$\boldsymbol{H}(v_l) = a_l \sum_{-\infty}^{\infty} \omega(v_l - s_k)\boldsymbol{G}'(s_k), \quad l = 1,2,\dots,N_r \,. \qquad \text{Equation 106}$$

Two new objects are introduced in Equation 106: the function ω and N_r constants a_l. Readers familiar with signal processing theory will recognize Equation 106 as a linear filter. Although the summation is infinite, it will be shown below that it can be truncated to include just a few points beyond the shooting interval.

The function ω has a bell shape with a width $\sigma S/N$, as shown schematically in Figure 141. The constants a_l are defined by the equations

$$a_l \sum_{-\infty}^{\infty} \omega(v_l - s_k) = 1, \quad l = 1,2,\dots,N_r \, .$$

These normalization conditions guarantee that the color levels in the reduced sequence images $\boldsymbol{H}$ will be in the same range as in $\boldsymbol{G'}$. If the function $\omega(u)$ is normalized so that

$$\int_{-\infty}^{\infty} \omega(u)du = \frac{S}{N},$$

then the constants a_l will be approximately equal to 1.

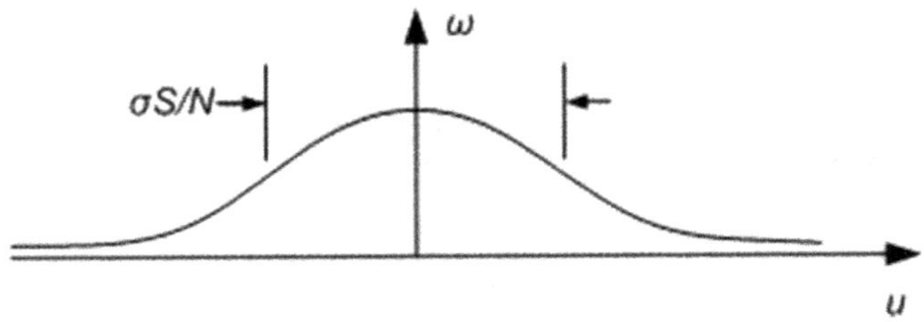

Figure 141: The function ω.

Due to the bell shape of the function ω, the sum in Equation 106 can be truncated to a finite number of terms. Let

$$M = \left\{ \frac{N + \sigma}{2} \right\}.$$

Equation 106 can be approximated by

$$\boldsymbol{H}(v_l) \approx a_l \sum_{-M}^{M} \omega(v_l - s_k)\boldsymbol{G}'(s_k), \quad l = 1,2,\dots,N_r \, . \qquad \textit{Equation 107}$$

The filter defined by Equation 106 (or Equation 107) mixes the graphical content of σ sequential images of sequence $\boldsymbol{G'}$. Therefore, its effect is qualitatively the same as the effect of the emulated blur. The intercell blurring of the emulated and the over-resolved methods is replaced by mixing with images taken outside the original shooting interval. Such mixing avoids the uncontrolled increase in incremental parallax caused by intercell blurring.

It can be concluded from the present discussion that one should choose $\sigma = y$. However, the optimal value of σ depends on the analytic properties of the function ω. The optimization of σ in the general case is a subject for future research.

A real example of the blurring effect is shown in Figure 142[5]. In this case, the length of the reduced sequence was 20 images, and the length of the original sequence was 40 images, corresponding to $y = 2$. The center images are displayed before and after the filtering, and the blurring effect is noticeable. Note that the blurring increases with the distance from the 3D window plane, which in this case passes through the body center of the model.

[5] Courtesy of Pop3DArt

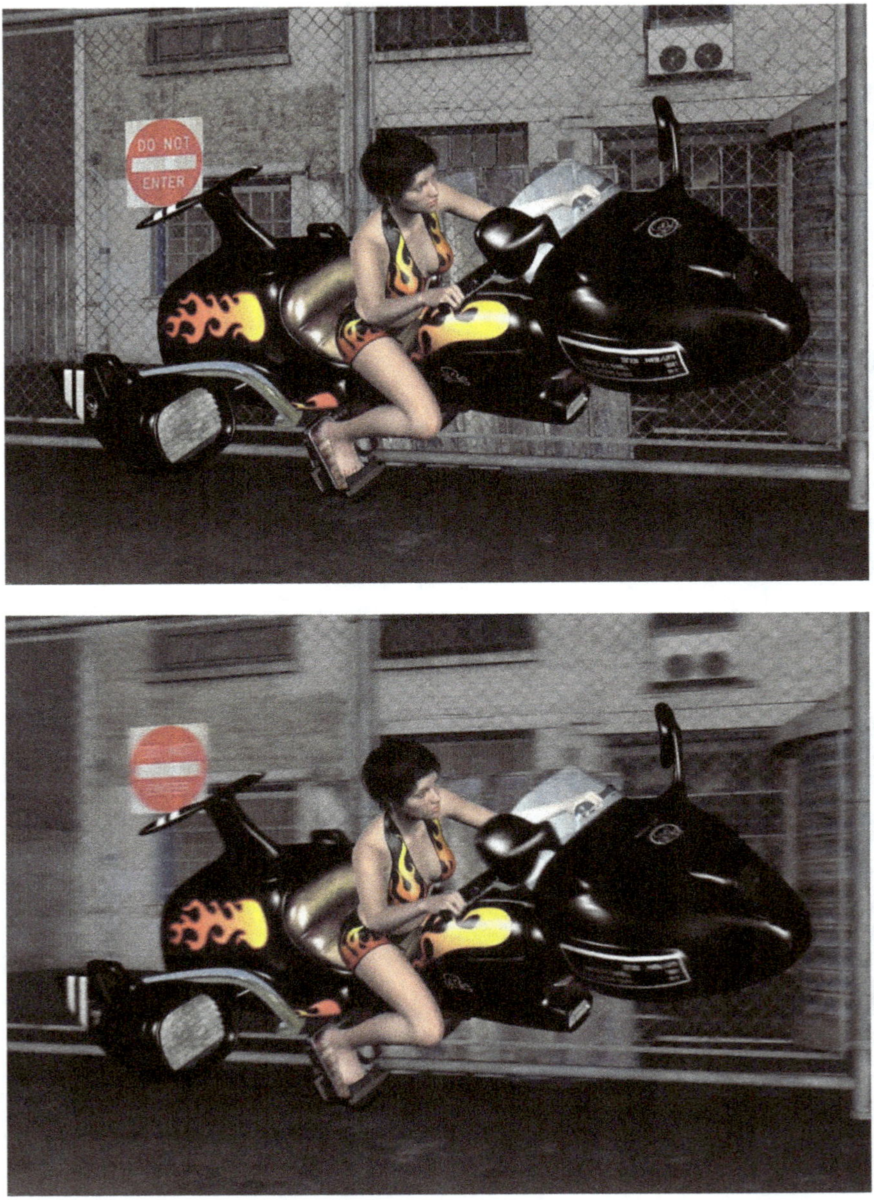

Figure 142: Reducing the sequence size by filtering. Top: image 20 from a 40-image sequence. Bottom: image 10 from a 20-image sequence produced from the 40-image sequence by filtering.

Chapter 10
Digital print

"But among the many lessons of history is that little of consequence is ever accomplished alone."

- David McCullough, in "The Great Bridge: The Epic Story of the Building of the Brooklyn Bridge"

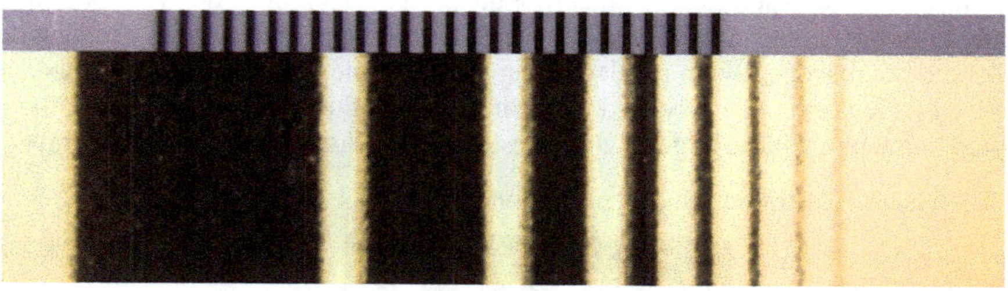

10.1 Printer characteristics for the lenticular application

The images in lenticular imaging are composed of grids. Therefore, the most important quality characteristics of the printer concern the printing of grids.

The first essential characteristic of the printer is the resolution. Since in most cases the picture resolution is limited by the print, the printer resolution determines the number of resolvable images, and therefore the depth that can be displayed.

The second characteristic is the ability of the printer to print regular grids, namely, produce an array of equally-spaced and equal-width lines. Departure from regularity produces visual artifacts and degrades the lenticular picture quality.

There are several printing technologies which are used for lenticular print, each having a different resolution and regularity.

10.2 Printer resolution

There are several ways to define a printer's resolution. In the present context, a printed pixel is regarded as being "resolved" if it is printed in the correct color and size. The relevant definition of printer resolution is the minimal resolved pixel size or the maximal printing density which gives resolved printed pixels.

There is a limit to the printing density which can be specified in the printer driver software. However, in most cases, this limit is much larger than the printer resolution. A good method to experimentally measure the printer pixel resolution is to print a sequence of black lines with decreasing width and measure their printed widths with a calibrated microscope. When pixels are resolved, the line width should be its pixel size divided by the printing density, and its color black. As the thickness decreases, the printed line width will be larger than this estimate, and its color will become gray.

To measure the resolution of a printer one can use the following pattern:

Figure 143: Pattern for measuring the printer resolution.

The line widths in pixels are powers of 2: 1, 2, 4, 8, 16, 32, 64, and 128. A magnified photo of the pattern printed with a 720dpi printer and dot density of 1440dpi is shown in Figure 144; the calibration grid at the top has a pitch of 0.2mm. The actual widths of the lines were measured with a microscope, and the results are shown in Table 27.

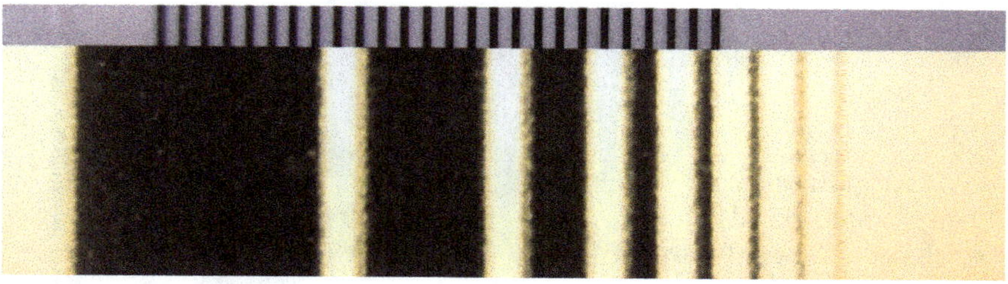

Figure 144: A photograph of the printed pattern.

Table 27: Results of printer resolution measurement.

line width (pixels)	measured line width (mm)	Calculated line width (mm)
1	0.056	0.018
2	0.058	0.035
4	0.08	0.071
8	0.17	0.141
16	0.31	0.282
32	0.59	0.564
64	1.12	1.129
128	2.32	2.258

It is evident that the thinnest line that this printer can print is about 0.06mm, which corresponds roughly to 450 pixels per inch, much lower than the maximally allowable printing density specification, which for this printer is 2880dpi.

10.3 Printer resolutions in other printing technologies

Researchers have addressed the problem of printer resolution quantification by studying the edge sharpness that the printer can produce (Lee & Winslow, 1993), (Streefland, 2006), (Christiansson & Blom, 2012). The edge sharpness is a measure of the transition width from the printed to the non-printed area. The minimal line width is approximately the sum of the left and the right edge sharpness values (which are, in general, different).

Figure 145 shows the magnified photographs of edges printed with three different printing technologies: inkjet, flexo, and offset (Streefland, 2006). The edge sharpness in the flexo and the offset print technologies is much better than in inkjet print.

200 x magnification of the edge of a printed black bar on 90 g/m² copy paper using a 300dpi inkjet printer.

200 x magnification of the edge of a printed bar-code on 140 g/m² white top kraft B flute corrugated board using a flexo printer.

200 x magnification of the edge of a printed bar on coated white kraft paper using an offset printer.

Figure 145: Edge sharpness in different printing methods (reproduced with permission from (Streefland, 2006)).

Another comparison of print characteristics is shown in Figure 146 (Christiansson & Blom, 2012).

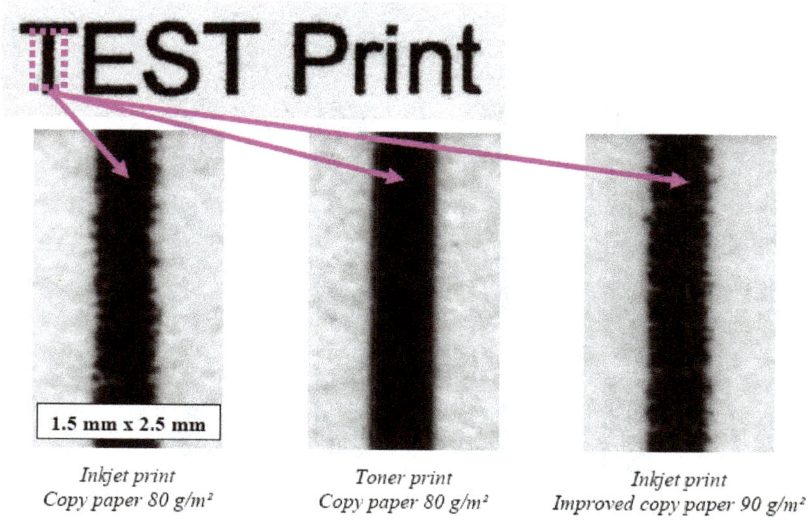

Figure 146: Edge sharpness in different printing methods (reproduced with permission from (Christiansson & Blom, 2012)).

10.4 Printer regularity

The printer regularity reflects its ability to print grids with constant pitch and constant column widths. In high-quality digital printers, the pitch variation is negligible. On the other hand, variations in column widths may be significant. These variations depend on the printing dot density.

Printers made by different manufacturers have different preferred printing densities in which the column width variations are minimized. For Epson printers, the preferred printing densities include 360, 720, and 1440dpi. For HP and Canon printers, the preferred printing densities include 300, 600, and 1200dpi. The preferred printing densities arise from the algorithms used for halftoning (Halftone, 2017) and the print head nozzles grid.

A pattern made of a grid of black lines may be used to illustrate the irregularities which appear in printed grids. In the example below, the pitch of the grid is 18 pixels, and the width of the black lines is 6 pixels (1/3 of the pitch). The image of this grid is shown in Figure 147. This image was printed on an Epson Stylus Pro 7900 printer with two densities: 720dpi (preferred) and 630dpi (non-preferred). Photos of the printed grids are shown in Figure 148 and Figure 149 (with different scales). The 720dpi grid is perfectly regular. In the 630dpi grid, it is possible to notice a slight periodic irregularity. Inspection under a microscope reveals that in this grid every seventh column is slightly thinner than the rest.

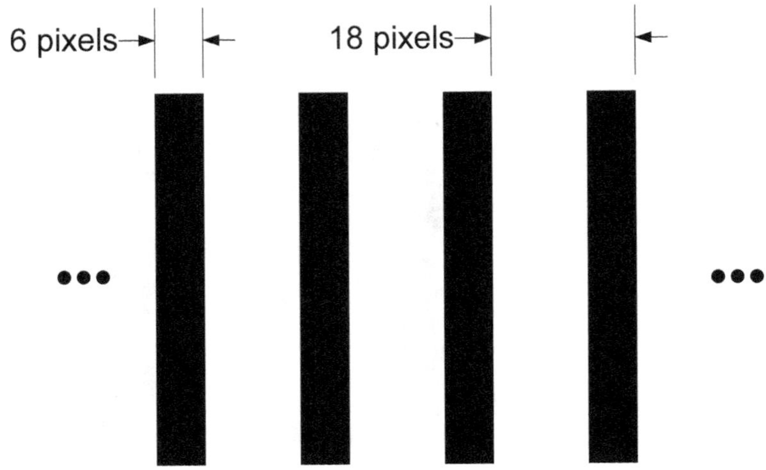

Figure 147: Image of the test grid.

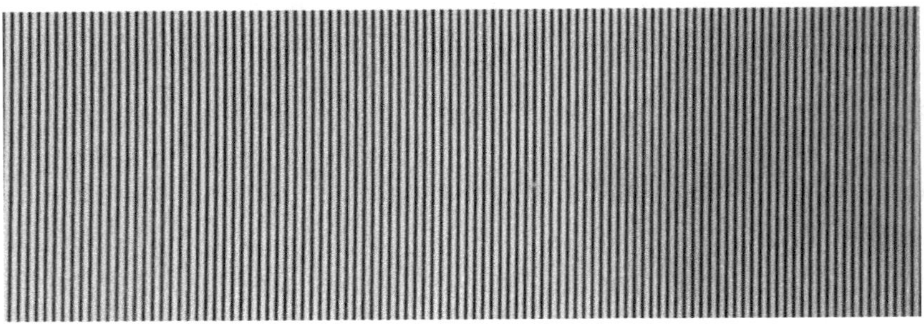

Figure 148: Photo of the 720dpi printed grid.

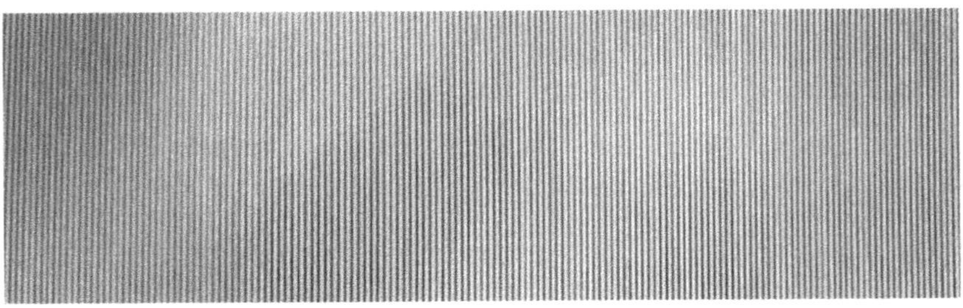

Figure 149: Photo of the 630dpi printed grid.

Although this irregularity is barely noticeable in the photo, it may be quite conspicuous in a lenticular picture. To illustrate the irregularity, the printed pattern cell size was matched to the optical grid pitch of a 40lpi lens, and a lenticular picture was made. Ideally, such picture should display a uniform color, either black or white, depending on the viewing angle.

A photo of the picture printed with 720dpi density is shown in Figure 150, and it displays a solid gray color with slight undulations. These undulations are due to rasterization effects. The picture with the 630dpi grid reveals a conspicuous artifact, as shown in Figure 151.

Figure 150: A lenticular picture of a grid pattern printed with 720dpi density.

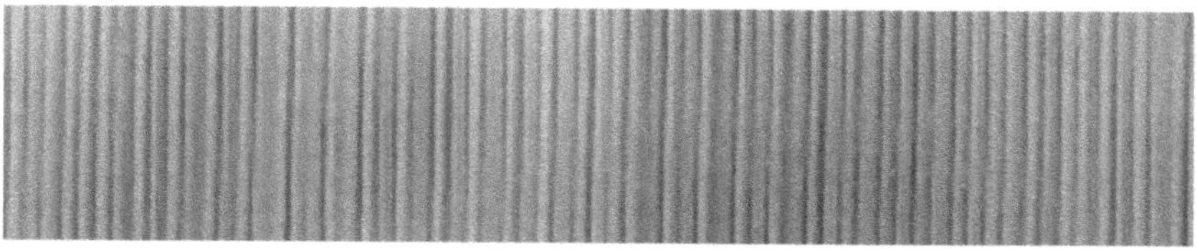

Figure 151: A lenticular picture of the same grid pattern as in Figure 150 printed with 630dpi density.

Both photos in Figure 150 and Figure 151 are shown in the same scale. They show a region approximately 19cm wide.

Epson inkjet printers' regularity was also studied by Gardner for some earlier models (Gardner, 2016).

Chapter 11
Optical resolution analysis

"...it is remarkable how much could be accomplished if you didn't care who received the credit."

- David McCullough, in "Truman"

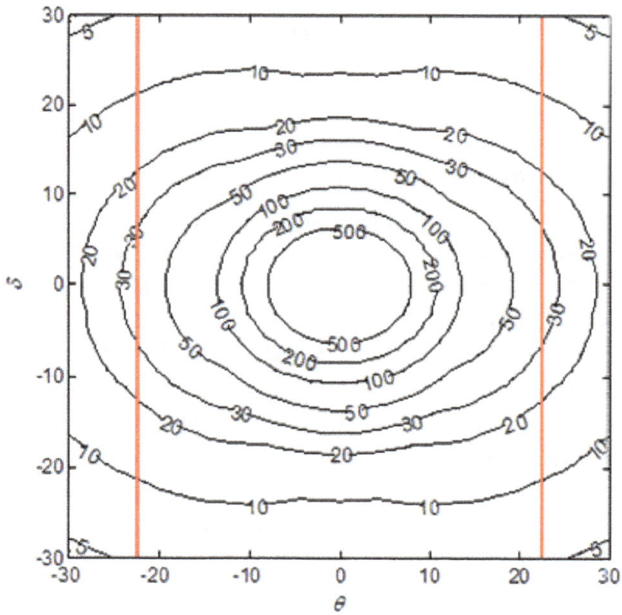

11.1 Mechanisms affecting optical resolution

The visible zone of a single lenticule has the form of a narrow column. The height of this column extends to the whole height of the lenticule, but its width is much smaller than the lenticule width. The visible zone hardly changes with vertical movement of the observer, but it translates horizontally with their horizontal movement.

There are four primary optical mechanisms which contribute to the visible zone width of a single lenticule:

1. Geometrical aberrations,
2. Chromatic aberrations,
3. Defocus,
4. Finite pupil size of the observer.

In the absence of these mechanisms, the visible zone of a lenticule is a straight line of zero width. Geometrical aberrations are deviations from the paraxial optics approximation (Jenkins & White, 1957). Most of this chapter will be devoted to the study of the geometrical aberrations on lenticule optics. The other mechanisms will be briefly discussed.

11.2 Ray tracing methodology

The main tool for understanding the topology of the visible zones is ray tracing. This method is used extensively in optical design to evaluate the performance of lens and optical systems. In ray tracing, we assume that light propagates as rays, which are reflected and refracted when passing from one medium to the other according to Snell's laws. These laws allow the computation of the trajectory of any ray as it propagates in an optical system.

When a ray encounters an interface between two dielectric media with refraction indices n_1 and n_2, it splits into two rays: refracted and reflected. All three rays remain in the so-called "incidence plane," which is the plane defined by the incident ray and the normal to the interface at the point of incidence. The three rays form three angles, α, β, and γ, as illustrated in Figure 152, and the angles α and γ are called the "incidence angle" and "refraction angle" respectively.

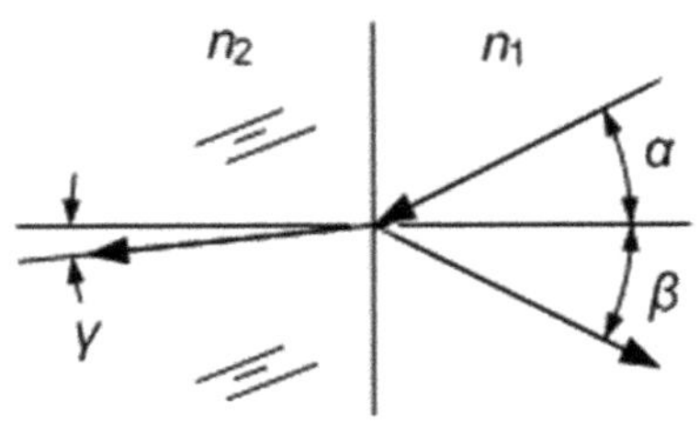

Figure 152: Snell's law.

Snell's law states the following:

$$\alpha = \beta \ ,$$

and

$$n_1 \sin \alpha = n_2 \sin \gamma \ .$$

The energy of the incoming ray is split between the reflected and the refracted rays in a manner described by other laws called "Fresnel laws." In the present application, the intensity of reflected rays is small, and they will be neglected. Also, in the present analysis, the incoming ray travels in air, for which n_1 is 1. Therefore, the following expression can be used for the refraction angle:

$$\sin \gamma = \frac{\sin \alpha}{n_2} \ . \qquad \textit{Equation 108}$$

The straightforward procedure to map the visible zone would be to consider all rays emitted by the printed picture and determine for each one whether it hits the observer's eye pupil. Such rays and points which emit them are considered "visible." The set of all visible points make up the visible zone.

The method of tracing rays from the light source to the eye is called "forward tracing," and is extremely inefficient. Only a tiny fraction of the rays emitted by the printed picture enters the observer's pupil, and ray-tracing of non-visible rays is a waste of computing time. Snell's laws are reciprocal, in the sense that any ray trajectory is valid in both forward and reverse directions. Therefore, instead of considering rays emitted from the printed image one may consider the observer's pupil hypothetically as a light source and trace the rays emitted by it instead of the rays emitted by the printed picture. Such a light source will illuminate a narrow column on the lenticule backplane, which is the visible zone. To a good approximation, the center of the visible zone coincides with the axial line shown as a broken backplane line in Figure 5. Since in this method only the visible rays are traced, it is much more efficient computationally. This method is called "reverse tracing," and is also used in the rendering of computer-generated 3D scenes. The concepts introduced in Figure 5 will be used in the ray-tracing analysis.

11.3 Ray tracing in the imaging plane

11.3.1 Ray tracing analytics in the imaging plane

The visible zone width in the imaging plane can be estimated by the analysis of the imaging rays, which is also two-dimensional. Ray tracing is used to determine the point ***H***, which is the intersection of a given imaging ray with the lenticule backplane. The coordinates of this point will be computed using the imaging plane diagram shown in Figure 153. This figure also shows the coordinate system used for the calculation. The origin of this system is defined as the axial point ***A***.

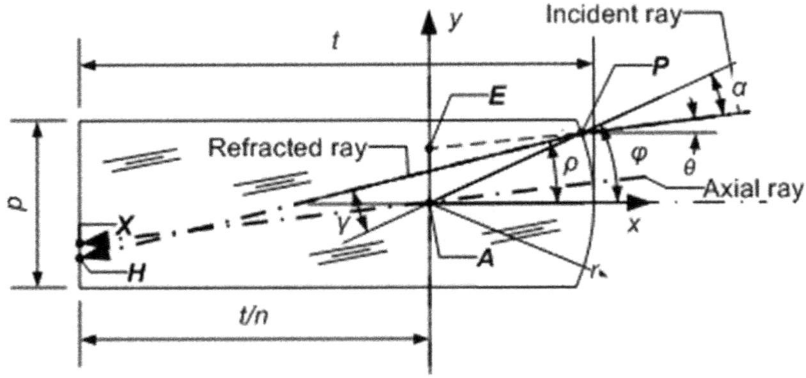

Figure 153: Imaging rays ray tracing.

Let us consider an incident ray with an incidence angle θ, which intersects the front surface of the lenticule at point ***P***. Let its continuation intersect the y-axis at point E. The equation of this ray is

$$y = x\tan\theta + E_y \ . \qquad \textit{Equation 109}$$

The equation of the lenticule front curve is

$$x^2 + y^2 = r^2 \ . \qquad \textit{Equation 110}$$

From Equation 109 and Equation 110 it follows that the x-coordinate of point P satisfies:

$$P_x^{\ 2} + \left(P_x \tan\theta + E_y\right)^2 = r^2 \ . \qquad \textit{Equation 111}$$

This quadratic equation yields two values of P_x, one positive and one negative, corresponding to the two intersection points of the ray with the circle described in Equation 110. The relevant intersection point is where the arc marks the curved surface of the lenticule, which is entirely on the positive side of the x-axis. Therefore, from the two values of P_x which solve Equation 111, the relevant one is positive. Once P_x is determined, P_y can be computed using Equation 109. The angle φ is derived from

$$\tan\varphi = \frac{P_y}{P_x} \ . \qquad \textit{Equation 112}$$

From elementary geometry, it follows that

$$\alpha = \varphi - \theta ,$$

and

$$\rho = \varphi - \gamma \ . \qquad \textit{Equation 113}$$

Using Snell's law (Equation 108) one gets

$$\rho = \varphi - \frac{\sin(\varphi - \theta)}{n} \; . \qquad \text{Equation 114}$$

The equation of the refracted ray is

$$y = x \tan \rho + Q \; . \qquad \text{Equation 115}$$

Since the point $\boldsymbol{P}$ is on the refracted ray, Q is given by

$$\boldsymbol{Q} = P_y - P_x \tan \rho \; ,$$

so, the y coordinate of point $\boldsymbol{H}$ is given by

$$\boldsymbol{H}_y = -\left(\frac{t}{n} + P_x\right) \tan \rho + P_y \; . \qquad \text{Equation 116}$$

This is the final result. Figure 153 also displays the axial ray which is parallel to the incident ray. This ray intersects the backplane at point $\boldsymbol{X}$. Within the paraxial optical approximation, points $\boldsymbol{H}$ and $\boldsymbol{X}$ coincide. However, ray tracing reveals that they do not, and allows the deviation to be computed. The y coordinate of point $\boldsymbol{X}$ is given by

$$X_y = -\frac{t}{n} \tan \theta \; . \qquad \text{Equation 117}$$

11.3.2 Imaging rays bundles

All imaging rays stay in the viewing imaging plane, and therefore all of their intersection points with the backplane lie on a straight line, which is the intersection of these planes. Imaging rays emerging from a given observation point will converge to a narrow interval on this line, which is the intersection of the lenticule viewing zone with the imaging plane. To evaluate the center and the width of this zone, one needs to trace a group of rays (typically 100) which emerge from the observation point and uniformly cover the lenticule aperture. Such a group of rays is called a "bundle".

The rays in the bundle before refraction are not parallel except for an observer at infinity. However, for all practical viewing scenarios which occur in lenticular imaging they are very nearly so. A parallel ray bundle is characterized by a single parameter only: the incidence angle. This fact greatly simplifies the presentation, and from now on we will deal with parallel ray bundles only. All dimensions will be normalized by the lenticule pitch p. By doing this, the number of parameters which define a given ray-tracing scenario is reduced to two only: the bundle angle of incidence and the lenticule relative pitch.

Let us start by considering the case of the normal incidence bundle shown in Figure 154. For normal incidence, the axial line is at the center of the backplane, and its footprint in the imaging plane is denoted in Figure 154 by point $\boldsymbol{X}$.

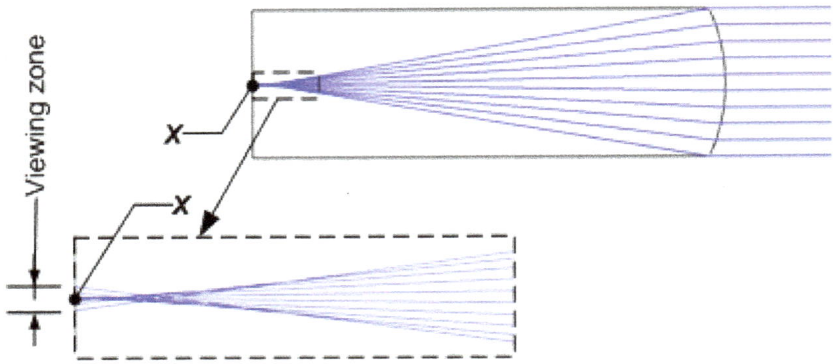

Figure 154: Imaging rays bundle at normal incidence.

A magnification of the focal region reveals the fact that not all rays converge to point ***X***. The intersections of the bundle rays with the lenticule backplane create a distribution of points. Such distribution for a 100-ray bundle at normal incidence is shown in the upper part of Figure 155. The horizontal axis shows the y coordinate normalized by the lenticule pitch p. The vertical axis is in the z-direction of the backplane and is in arbitrary units. The origin of the z-axis is chosen to be on the imaging plane. Since all bundle rays stay in this plane, the z coordinates of all intersection points are 0. The computation was done for a lenticule with a relative pitch of 0.5, and the vertical line is the distribution center, in this case also the center of the lenticule backplane.

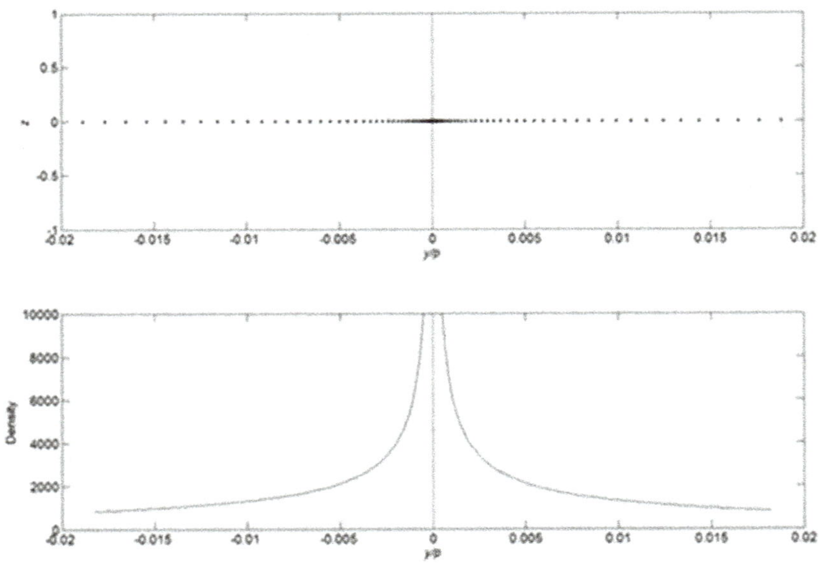

Figure 155: Intersection points of the normal incidence bundle with the lenticule backplane, R_p = 0.5.

Let us now repeat this analysis for the same scenario except that the incidence angle is taken to be 10°. The result of such ray tracing is shown in Figure 156, and the corresponding intersection point distribution is shown in the upper part of Figure 157. Naturally, it is not symmetrical; most rays still converge close to the axial point, but this time the scatter is much larger. The solid vertical line marks the center of the distribution, and the broken lines its effective boundaries (see below).

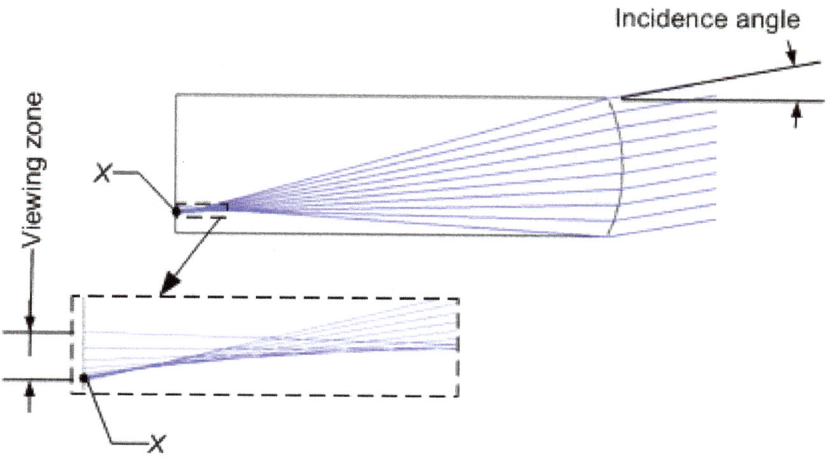

Figure 156: Imaging rays bundle for incidence angle of 10°.

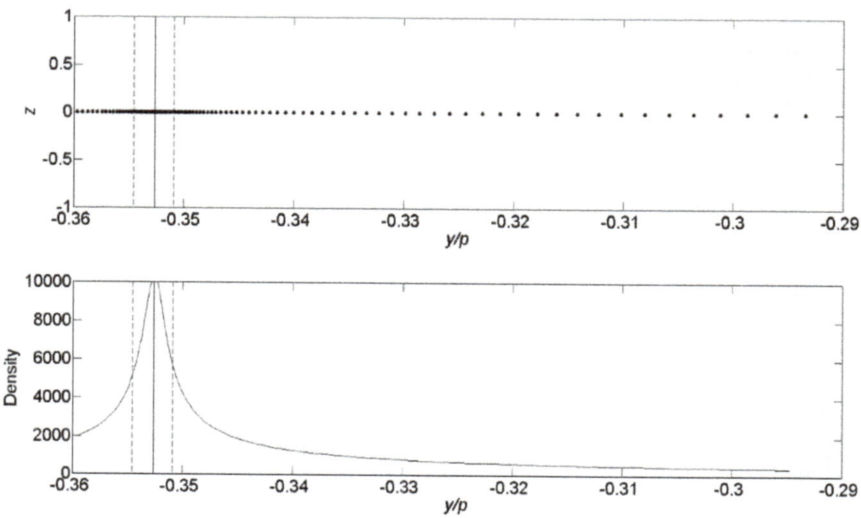

Figure 157: Intersection points of 10° incidence bundle with the lenticule backplane, R_p = 0.5.

11.3.3 The center and the boundaries of the visible zone

Strictly speaking, the visible zone is the whole interval between the two extreme intersection points of the bundle. Thus, for normal incidence, the viewing zone is the interval [-.02 .02], and for the 10° incidence, it is [-0.36 -0.29] (dimensions are normalized to p). However, its effective width is much smaller. Besides, its "center" is not necessarily the arithmetic mean value of the distribution.

The content which is displayed on the lens aperture is a composite sample of the printed content in the visible zone. Each point in this zone contributes to the composite sample with a different weight. It is claimed, without proof, that this weight is proportional to the point distribution density at the sampled point. This claim will be used to determine the effective center and edges of the visible zone.

The lower parts of Figure 155 and Figure 157 show the density values corresponding to the point distributions shown above them. The vertical solid line denotes the position of the maximal density, and the broken lines the points in which the density drops by a factor of 2 with respect to the maximal value. The solid line will be considered as the distribution center and the broken lines as the effective edges of the visible zone. The visible zone width is estimated as the distance between these two lines. Note that there are no broken lines in Figure 155 because in this case, the effective width is too small to be resolved by a bundle of 100 rays. This estimate of distribution width differs from the standard width characterization using the statistical standard deviation.

To sum up, a distribution of intersection points can be characterized by two parameters: center and width. These parameters will be considered as the effective center and width of the visible zone. Figure 158 and Figure 159 show several distributions in the imaging plane with their centers and widths. The relative pitches in these figures are 0.5 and 1 respectively.

11.3.4 The significance of the acceptance angle interval

The qualitative behavior of the viewing zone parameters can be explained with the help of the acceptance angle that was introduced in Figure 3. Figure 160 shows a case in which the horizontal viewing direction is inside the lenticule acceptance angle interval. In such a case, all rays will converge to a relatively small interval around the point $\boldsymbol{X}$ which is the intersection of the axial ray with the backplane and marks the footprint of the corresponding axial line.

When the horizontal viewing direction is outside the acceptance angle interval, the axial rays are refracted, and the intersection points of the incoming rays with the backplane become more dispersed. This is shown schematically in Figure 161. This figure also shows, for the sake of illustration, a hypothetical unrefracted axial ray and its corresponding intersection point $\boldsymbol{X}$ with the backplane. Such ray does not exist in reality.

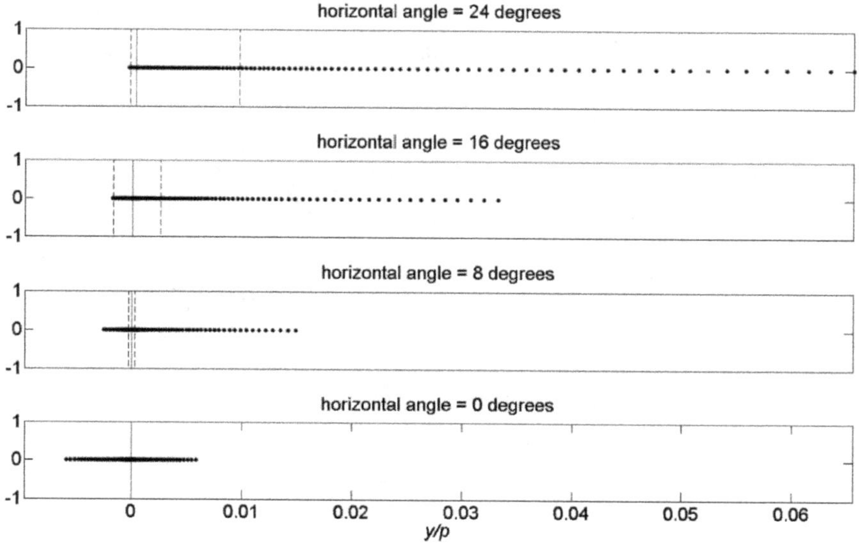

Figure 158: Backplane intersection distribution for parallel bundles, relative pitch 0.5.

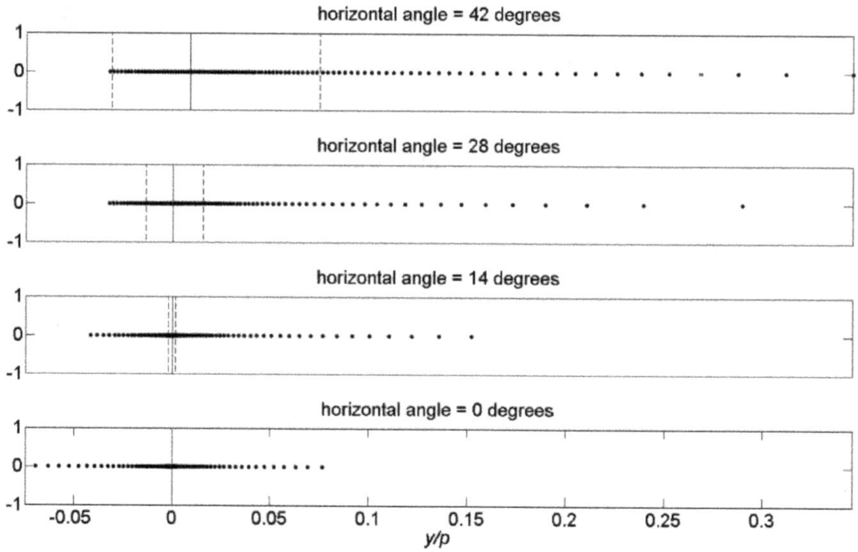

Figure 159: Backplane intersection distribution for the parallel bundle, relative pitch 1.

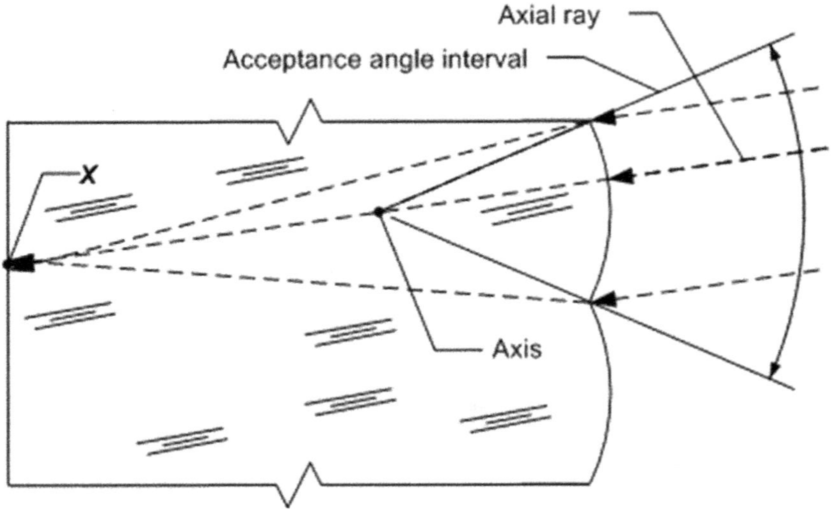

Figure 160: Horizontal viewing direction inside the lenticule acceptance angle interval.

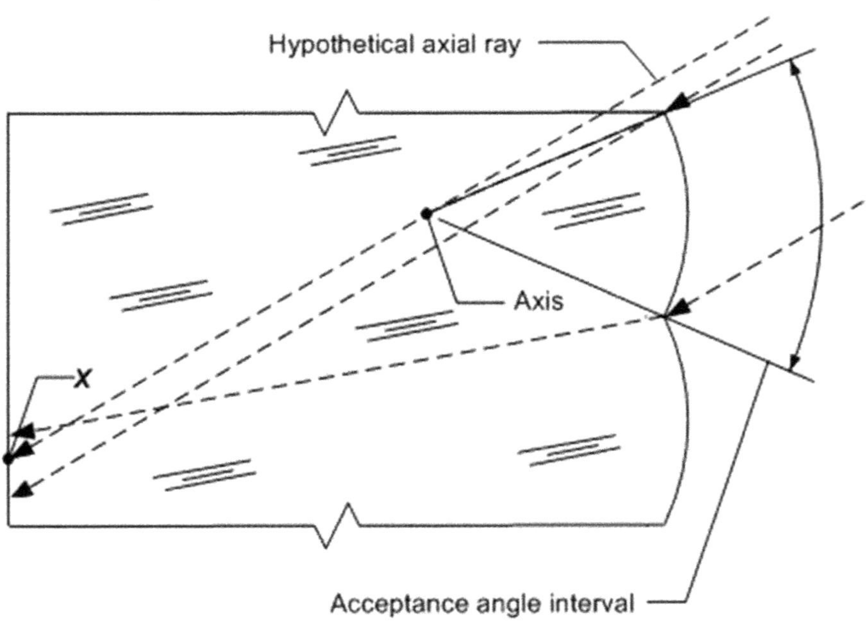

Figure 161: Viewing direction outside the lenticule acceptance angle interval.

11.3.5 Optical characteristics in the imaging plane

A lenticular sheet is an array of identical lenticules, each having its own viewing zone. Lenticular printing relies on having all viewing zones centered on the axial lines. Deviations of the viewing zones centers from the axial lines cause several adverse visual defects and in general worsens the picture quality. The quantity of interest is not the absolute value of the deviation, but rather its value compared to the visible zone width. So, the center displacement will be characterized by the ratio δc:

$$\delta c = \frac{\text{center deviation}}{\text{zone width}} .$$

The width of the viewing zone determines the number of lines that the lenticule can resolve. In the context of optical science, this is called the "number of resolvable lines." The number of resolvable lines will be denoted by N_o:

$$N_o = \frac{p}{\text{zone width}} . \qquad \textit{Equation 118}$$

This quantity is related to the optical resolution of the lenticule: higher value of N_o indicates higher optical resolution.

The quantities N_o and δc were computed for a range of horizontal angles values and two relative pitches: 0.5 and 1. The results are shown in Figure 162 and Figure 163. The vertical solid lines denote the boundaries of the acceptance angle interval. The upper plots demonstrate the significance of the acceptance angle. For angles within the acceptance angle interval, the relative center deviation is small and starts to rise sharply as the boundary of this interval is approached. This is caused by the disappearance of the axial ray beyond the acceptance angle interval as illustrated in Figure 161. The effect is more pronounced in Figure 162 (relative pitch 0.5) than in Figure 163 (relative pitch 1).

At zero incidence angle, the optical resolution is very high, practically infinite. As expected, the value of N_o drops sharply with the deviation of the incidence angle from normal. At 20°, the value of N_o is approximately 50 for both $R_p = 0.5$ and $R_p = 1$.

This paragraph concludes the analysis of the imaging plane, which was done in two dimensions. Quantitative data were presented on the optical resolution, and the deviation of the viewing zones centers from a regular grid. However, to see the whole picture, it is necessary to undertake analysis in three dimensions and investigate the effect of the skew rays.

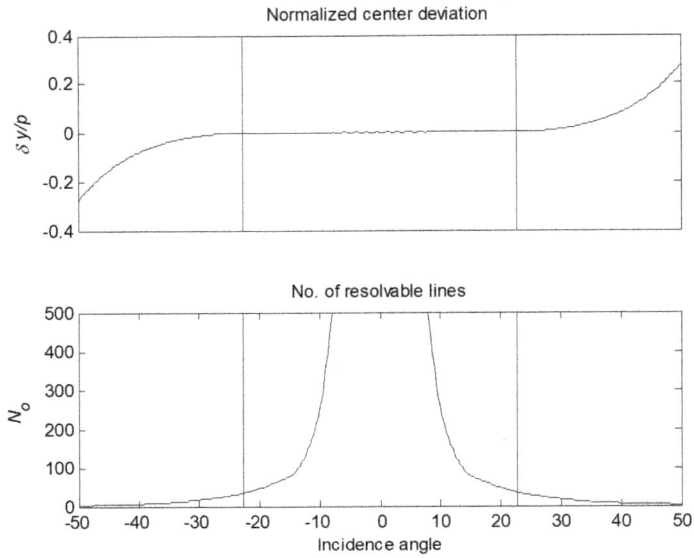

Figure 162: Optical characteristics in the imaging plane as a function of the incidence angle for relative pitch of 0.5.

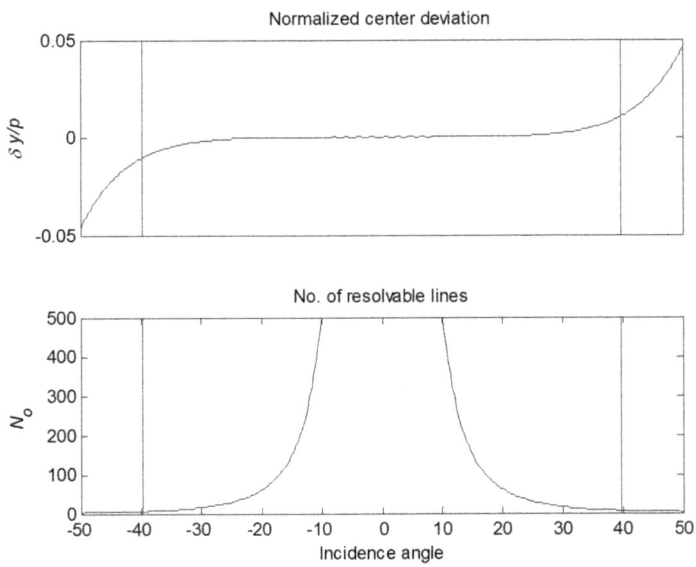

Figure 163: Optical characteristics in the imaging plane as a function of the incidence angle for relative pitch of 1.

11.4 Three-dimensional ray tracing analysis

11.4.1 Coordinate system and angles for three-dimensional ray tracing

The coordinate system for three-dimensional ray tracing analysis is defined in Figure 164. The origin of the system lies on the lens axis. The z-axis coincides with the lenticule axis, the x-axis is perpendicular to the lens backplane, and the y-axis is parallel to it. The origin is chosen so that the x-y plane contains the observation point (not shown in Figure 164).

An x-y plane is an imaging plane, and all axial planes contain the z-axis.

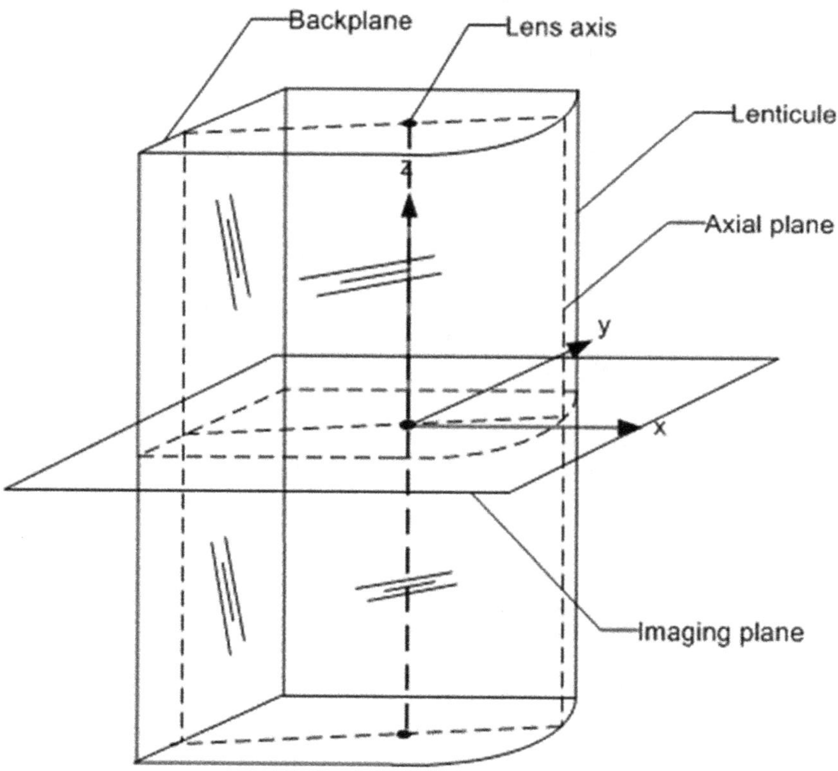

Figure 164: Coordinate system for ray tracing.

It is convenient to introduce viewing angles corresponding to a given observation point on the lenticular picture. These angles are defined in Figure 165. In practically all viewing scenarios R is much larger than the lenticule thickness t, and, accordingly, the lenticule is represented in Figure 165 by a thick line.

The observer's eye is considered as being located on an arc with a radius R and with a center at the lenticule. The position of the eye on this arc is given by the horizontal viewing angle θ. The angle between the vector pointing from the eye to the observation point and the x-y plane is called the vertical viewing angle and is denoted in Figure 165 by δ. The height h of the observation point above the x-y plane is given by

$$h = R\tan\delta \ . \qquad \text{Equation 119}$$

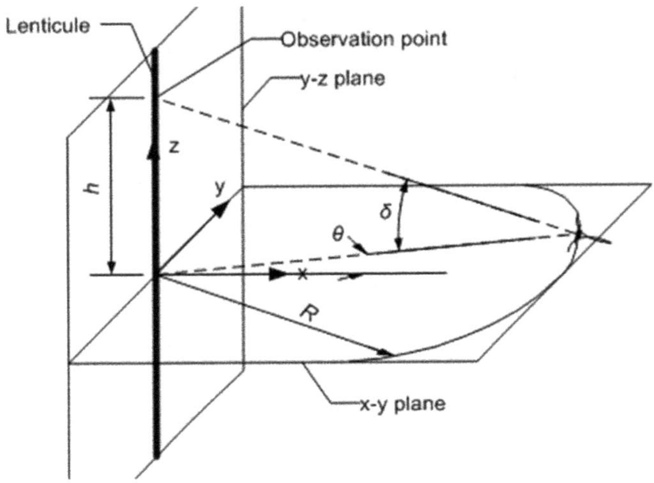

Figure 165: Viewing angles.

11.4.2 Ray tracing algorithm

The diagram shown in Figure 166 is used to trace rays in a general three-dimensional scenario.

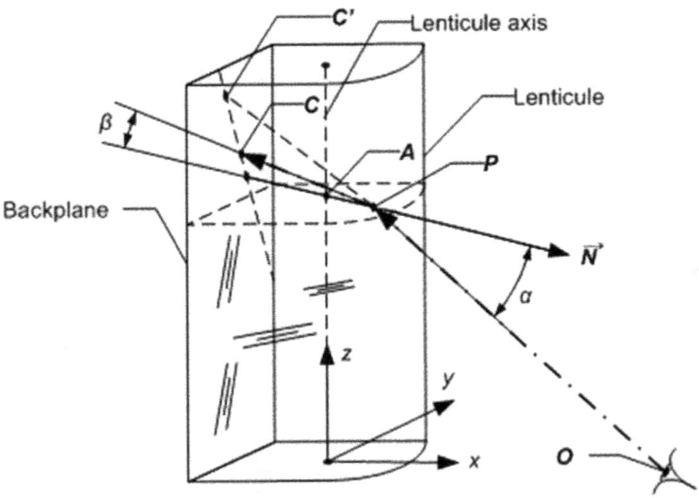

Figure 166: Diagram for ray tracing of general rays

Consider a general ray originating from the center of the observer's pupil ***O*** and directed towards a given point ***C'***. In Figure 166 this point is shown to be on the lenticule backplane, but this is just for the sake of illustration; it can be located anywhere. The parametric equation of all points ***q*** on this ray is

$$\boldsymbol{q}(v) = \boldsymbol{O} + v\cdot(\boldsymbol{C'} - \boldsymbol{O}). \quad \textit{Equation 120}$$

Let ***P*** be the intersection point of this line with the curved surface. The components of ***P*** satisfy the equation

$$P_x^2 + P_y^2 = r^2 \quad \textit{Equation 121}$$

Let us denote by v_0 the value of v corresponding to the intersection point ***P***. By combining Equation 120 and Equation 121 one gets

$$[O_x + v_0(C'_x - O_x)]^2 + [O_y + v_0(C'_y - O_y)]^2 = r^2 . \quad \textit{Equation 122}$$

This is a quadratic equation from which v_0 can be determined. The point ***P*** is then given by

$$\boldsymbol{P} = \boldsymbol{O} + v_0\cdot(\boldsymbol{C'} - \boldsymbol{O}). \quad \textit{Equation 123}$$

There are in general two roots for Equation 122, and for this treatment the root that gives a positive P_x should be chosen, for the same reasons that the positive root was chosen for Equation 110.

Let ***N*** be the normal to the curved surface at ***P***. This normal intersects the lenticule axis in point ***A***:

$$\boldsymbol{A} = [0, 0, P_z] . \quad \textit{Equation 124}$$

For any ray which is not the main ray the points ***O***, ***P***, and ***A*** do not lie on the same line. For any such ray, the normal ***N*** is given by

$$\boldsymbol{N} = (\boldsymbol{A} - \boldsymbol{P}) \times (\boldsymbol{P} - \boldsymbol{O}), \quad \textit{Equation 125}$$

where '**x**' stands for vector product.

According to Snell's law (Figure 152), the refracted ray is contained in the plane defined by the incident ray and the normal ***N***. All points ***x*** belonging to the refraction plane satisfy the equation

$$\boldsymbol{N}\bullet\boldsymbol{x} + d = 0.$$

Here the symbol "•" stands for scalar product. Since the point ***A*** lies on this plane

$$N_z A_z + d = 0$$

so that

$d = - N_z A_z$.

The x coordinate of point $\boldsymbol{C}$ is $-t/n$, and it lies on the refraction plane. Therefore, its components satisfy the equation

$$N_y C_y + N_z C_z = N_z B_z + N_x \cdot \frac{t}{n} . \qquad \textit{Equation 126}$$

The incidence angle α can be computed from the equation

$$\cos\alpha = \frac{(\boldsymbol{P}-\boldsymbol{O})\bullet \boldsymbol{N}}{|\boldsymbol{P}-\boldsymbol{O}|\cdot|\boldsymbol{N}|} , \qquad \textit{Equation 127}$$

where $|\boldsymbol{W}|$ stands for the Cartesian length of vector $\boldsymbol{W}$. The angle β is given by Snell's law:

$$\sin\beta = \frac{\sin\alpha}{n} . \qquad \textit{Equation 128}$$

The cosine of this angle can also be written as

$$\cos\beta = \frac{(\boldsymbol{P}-\boldsymbol{C})\bullet \boldsymbol{N}}{|\boldsymbol{P}-\boldsymbol{C}|\cdot|\boldsymbol{N}|} . \qquad \textit{Equation 129}$$

It is now possible to specify the algorithm for the calculation of point $\boldsymbol{C}$, which is valid for all incident rays which are not the main ray.

Given: $\boldsymbol{O}$, $\boldsymbol{C'}$, t, n:

1. Calculate point $\boldsymbol{P}$ (Equation 123) by solving for v_0 (Equation 122)
2. Calculate point A (Equation 124)
3. Calculate normal $\boldsymbol{N}$ (Equation 125)
4. Calculate angle α (Equation 127)
5. Calculate angle β (Equation 128)
6. Solve for C_y and C_z (Equation 126 and Equation 129)

The last step requires the solution of a non-linear equation. This equation will, in general, have two solutions corresponding to refracted rays with angles $+\beta$ and $-\beta$. Only one of the solutions is optically correct. The correct solution yields the point $\boldsymbol{C}$ closest to the axial line corresponding to point $\boldsymbol{O}$.

11.4.3 The three-dimensional rays bundle

A bundle of rays was used to investigate the characteristics of the visible zone in the imaging plane. The same approach will be used in the three-dimensional analysis. The bundle of incident rays will be a group of rays originating from the observation point with (unrefracted) continuations that cross a certain interval called "target interval" at an array of equidistant points. The target interval is horizontal, and it crosses the edges of the lenticule curved surface. Such a bundle of rays is shown in Figure 167. For clarity, only five rays are shown; in the analysis, a much larger number of rays is used, typically 100. The refracted rays converge on the backplane near the axial line. This definition is a generalization of the bundle concept used for the two-dimensional analysis. The five black dots on the rays denote the intersection points with the lenticule curved surface. The refracted rays do not cross, in general, the target interval, except at its edges (which lie on the lenticule curved surface).

As in the two-dimensional analysis, it will be convenient to deal with parallel bundles, because they can be fully characterized by two parameters only: the horizontal and the vertical viewing angles θ and δ which were defined in Figure 165. Again, parallel bundles correspond to an observer at infinity; however, in all practical viewing scenarios the bundles are nearly parallel, so parallel bundles are an excellent approximation to real bundles. With parallel bundles, the analysis results can be presented in a very compact and clean form. In fact, all results depend on one physical parameter only: the lens relative pitch.

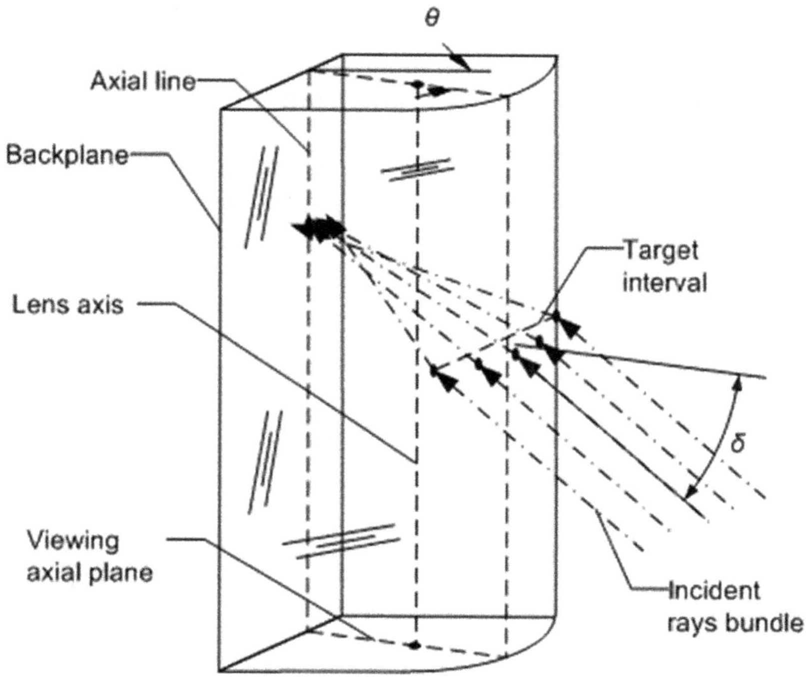

Figure 167: Incident ray bundle used for ray tracing analysis.

11.4.4 Backplane point distributions for a zero horizontal viewing angle

The bundle ray-tracing maps the target interval to a curve on the lenticule backplane. The imaging plane analysis of paragraph 10.3.3 corresponds to a zero vertical angle. In such a case, the mapped curve is a straight line. However, for a non-zero vertical viewing angle, the curve extends in both dimensions. In Figure 168 and Figure 169 such curves are shown in the y-z plane for relative pitches of 0.5 and 1 respectively. The z-coordinates are shown relative to the minimum of the distribution.

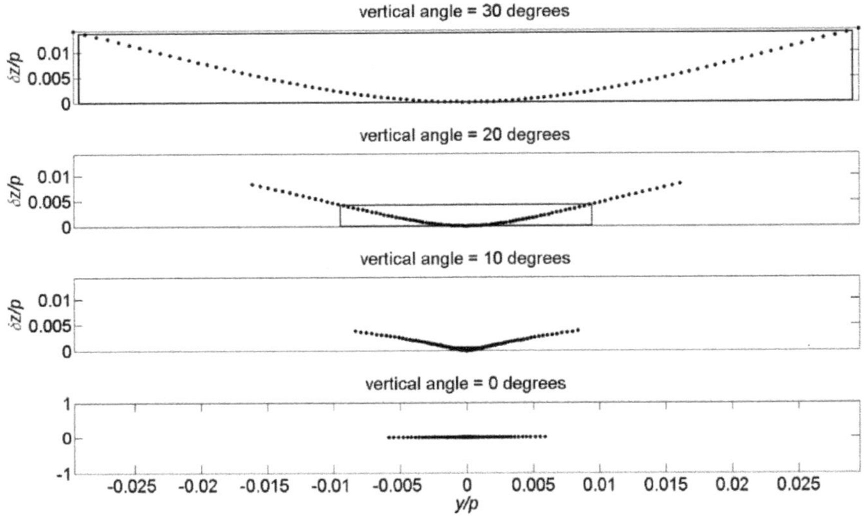

Figure 168: Point distributions of observation point bundles for different vertical viewing angles. Relative pitch 0.5.

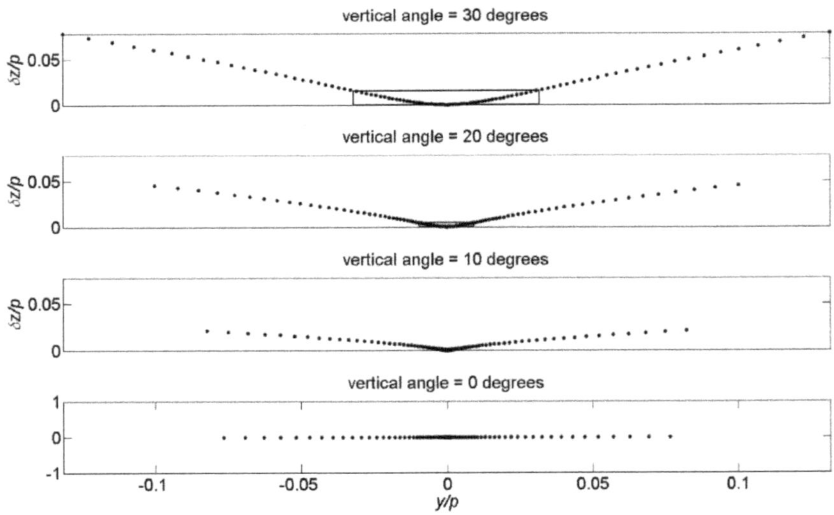

Figure 169: Point distributions of observation point bundles for different vertical viewing angles. Relative pitch 1.

The width and the height of these two-dimensional distributions are estimated as follows. For the width estimation, the method in paragraph 11.3.3 is used as applied to the y coordinates of the distribution. For the height estimation, the values of the z coordinates of the curve (relative to the minimum) for the computed y-axis limits are used. This defines a rectangle which marks the effective extension of the curve, which is a generalization of the distribution width concept used to characterize point distributions in the imaging plane. Such rectangles are drawn for each curve in Figure 168 and Figure 169, except for the zero-angle curve. Due to the symmetry of this scenario, the centers are always at $y = 0$.

The rectangles in Figure 168 and Figure 169 mark the effective boundaries of the respective target interval mappings. The fact that these rectangles have a finite height indicates that the optical resolution is limited in the vertical dimension too. However, it is seen from Figure 168 and Figure 169 that the height of the rectangles is much smaller than their width.

11.4.5 Optical resolution as a function of the vertical angle for a zero horizontal angle

The concept of the number of resolvable lines was introduced in 11.3.5 to characterize the optical resolution. The same concept is used here to compute the optical resolution as a function of the vertical viewing angle. The viewing zone rectangle width is used to estimate the number of resolvable lines. The results are shown in Figure 170 for the two values of the numerical aperture, 0.5 and 1.

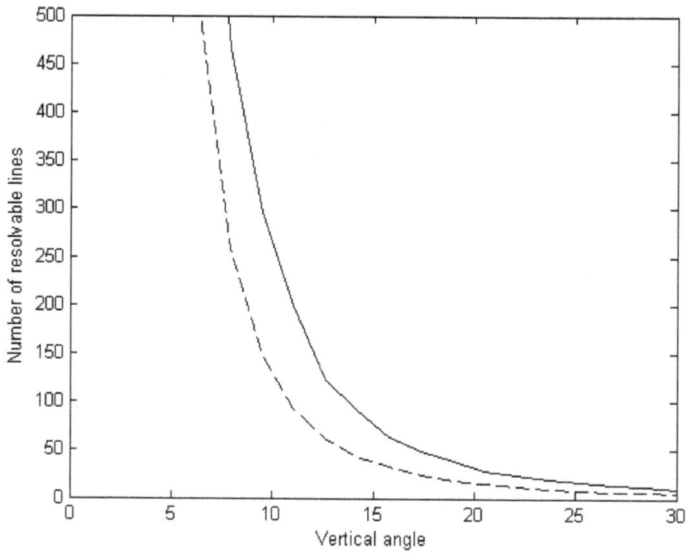

Figure 170: Number of resolvable lines as a function of the vertical angle. Broken curve: relative pitch 0.5, solid curve: relative pitch 1.

11.4.6 Optical resolution in the angle plane

Let us turn now to the general case of any combination of horizontal and vertical angles θ and δ. The results can be displayed as contour plots in the $\theta \times \delta$ plane. The computations were done for relative pitch values of 0.5 and 1. The results are shown in Figure 171, Figure 172, Figure 173, and Figure 174. The vertical lines mark the boundaries of the corresponding acceptance angle intervals.

Inspection of Figure 171 and Figure 173 reveals that the number of resolvable lines drops faster with the vertical angle than with the horizontal angle. This is more pronounced for a relative pitch of 0.5. In this respect, landscape pictures make better use of the optical resolution resources than portrait pictures.

The acceptance angle interval is manifested in the contour shapes of the normalized deviation shown in Figure 172 and Figure 174. The area within the acceptance angle interval has a small value of the normalized displacement, and it starts to rise sharply outside this area. It can also be seen that the deviation has a weak dependence on the vertical angle. This implies that the viewing zone columns are slightly curved, in particular beyond the acceptance angle interval.

11.5 Other mechanisms which affect the optical resolution

11.5.1 Introduction

Geometric aberrations are not the only mechanism affecting the optical resolution. From the additional mechanisms, three will be discussed here, which are probably the most important after the geometric aberrations.

Each mechanism (including the aberrations analyzed above) causes the observer's pupil to be mapped to a spot of finite size. The widths of these spots b_i will be estimated for centered viewing (zero horizontal and vertical angles) using analytical methods. The combined spot width b of the various mechanisms can be estimated by

$$b^2 = \sum_i b_i^2 \ .$$

Again, the more significant quantity in the present context is the limitation on the number of lines that can be resolved. This is given by

$$N_i = \frac{p}{b_i} \ ,$$

and the combined limitation on the number of lines is

$$N = \frac{p}{b} \ .$$

The resolution limiting mechanisms discussed here are listed in Table 28.

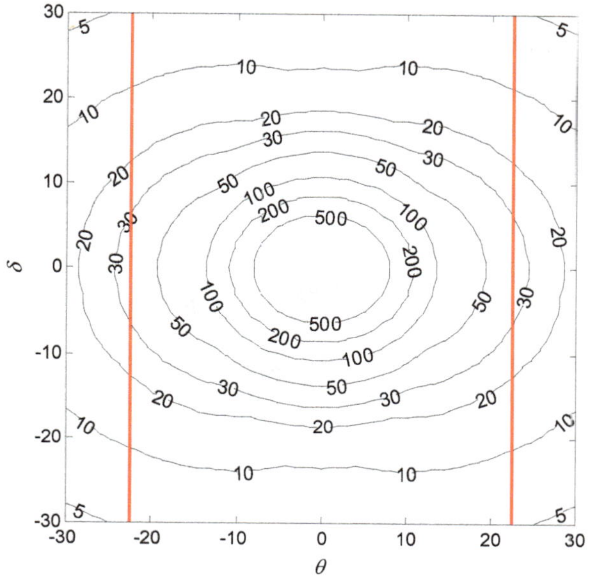

Figure 171: Number of resolvable lines in the θxδ plane. R_p = 0.5.

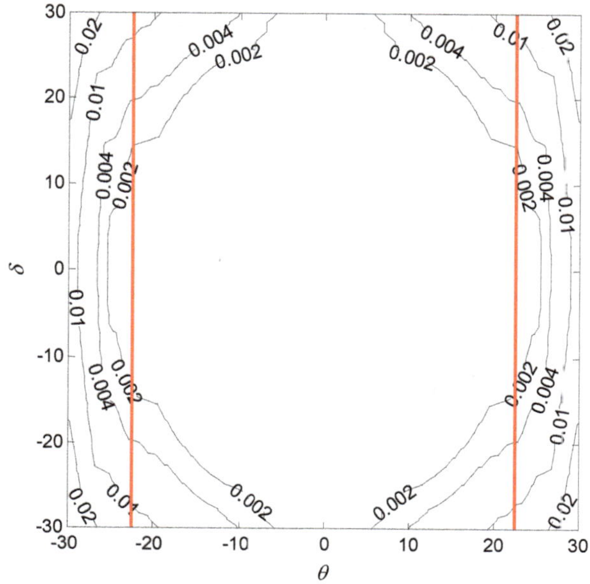

Figure 172: Normalized center deviation in the θxδ plane. R_p = 0.5.

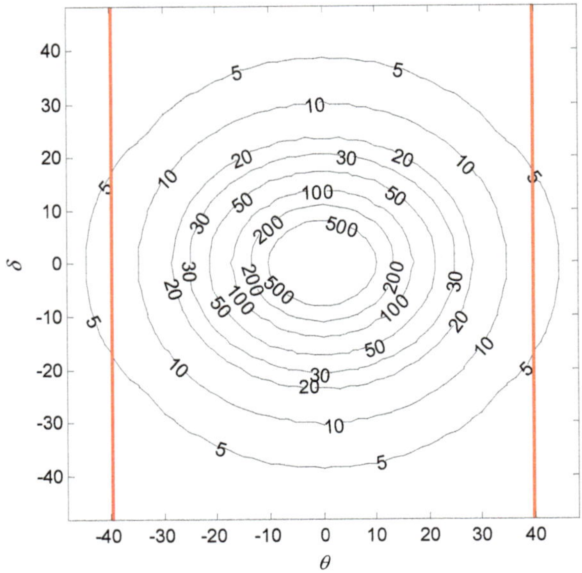

Figure 173: Number of resolvable lines in the θxδ plane. R_p = 1.

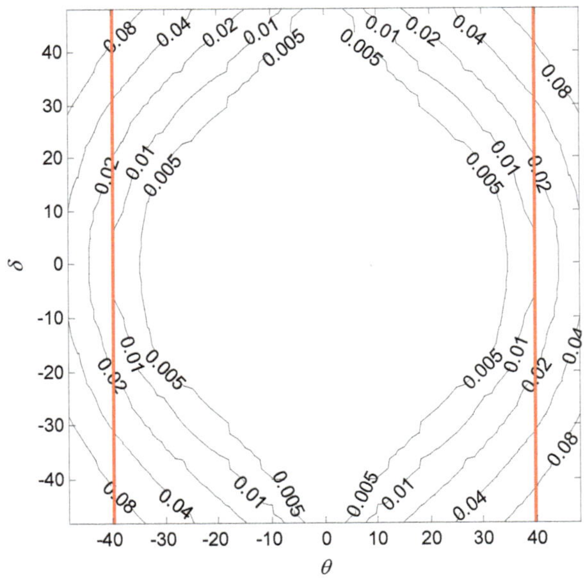

Figure 174: Normalized center deviation in the θxδ plane. R_p = 1.

Table 28: Resolution limiting mechanisms

Mechanism	Spot size
Deviation from focus	b_f
Finite pupil size	b_a
Chromatic aberration	b_c

11.5.2 Deviation from focus

In most cameras, there is a provision to "focus" the lens. This operation adjusts the position of the lens so that the sharpest image of the object of interest will appear on the sensor. If this operation is not carried out, the image will, in general, appear blurred, and can be described as being "out of focus."

There is no focusing mechanism in a lenticular picture. The "focus" of a lenticule is set to infinity. Therefore, b_f is zero only for an observer at infinity. For any finite viewing distance, the value of b_f is finite, and it increases when the observer approaches the lenticule. This is illustrated in Figure 175. In this figure, we have artificially extended the dielectric material of the lenticule beyond the focal plane to avoid the unnecessary complication of drawing refracted rays as they emerge from the lens. This action does not influence the discussion.

Figure 175 shows two imaging ray bundles, both created by a point light source. In one case, the light source is located at infinity, so all rays coming from it are parallel to each other. In the other case, the light source is located at a certain distance v, and the rays coming from it are divergent. The ray bundle that comes from infinity is focused at a distance f behind the lens surface, and the other one is focused at a certain distance s. The contribution b_f corresponding to the finite light source is the spread of its ray bundle at the back focal plane, as indicated in Figure 175.

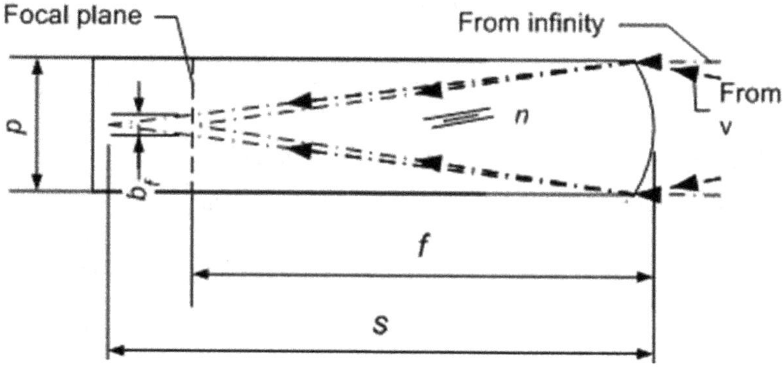

Figure 175: Loss of resolution due to deviation from the optimal viewing distance v.

The distances s and v are related to each other by Equation 1, where f is substituted by the lenticule thickness t:

$$\frac{n}{s}+\frac{1}{v}=\frac{n}{t}. \qquad \text{Equation 130}$$

Rearranging this equation yields:

$$s-t=\frac{st}{nv}. \qquad \text{Equation 131}$$

Using similarity relations in Figure 175, one gets

$$\frac{b_f}{s-t}=\frac{p}{s}, \qquad \text{Equation 132}$$

so that

$$b_f=\frac{t}{nv}p. \qquad \text{Equation 133}$$

The corresponding limitation N_f on the number of lines is

$$N_f=\frac{nv}{t}=R_p\frac{v}{p}.$$

11.5.3 Finite size of the viewer's pupil

The spot size b_a corresponding to a given pupil size a can be computed with the help of Figure 176.

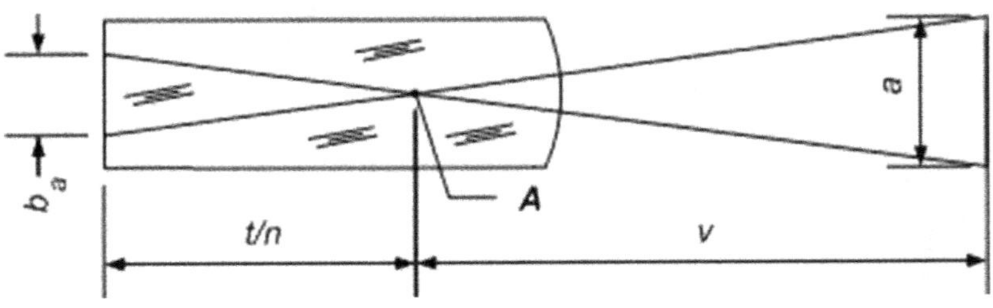

Figure 176: Spot size corresponding to finite pupil.

Using similarity relation between the two triangles we obtain

$$b_a=\frac{t}{nv}a. \qquad \text{Equation 134}$$

The pupil diameter size in humans varies between two extreme values: 1mm to 9mm. The dominant factor which determines its size is illumination, but other factors include age and gender. Taking a representative value of 6mm for typical estimations is common.

The largest value of p found in commercial lenses is 2.5mm. In most commercial sheets, the value of p is less than 1.5mm. Therefore, it is possible to neglect b_f with respect to b_a.

The limitation on the number of lines due to this mechanism is

$$N_a = \frac{nv}{t}\frac{p}{a} = R_p \frac{v}{a} \,. \qquad \textit{Equation 135}$$

The relative contribution of the finite pupil mechanism can be reduced by increasing the light intensity to that which the observer is exposed. High light intensity reduces the pupil's size, and correspondingly it's associated spot width.

11.5.4 Chromatic aberrations

Chromatic aberrations are caused by the fact that the refraction index n depends on the wavelength of light, a phenomenon called "Dispersion". By Equation 4, the lenticule thickness t depends on the value of n. This means that t is correct only for a certain wavelength. The light with other wavelengths will be out of focus. This may cause two effects:

1. Loss of optical resolution,
2. Color distortions.

From Equation 131 it can be derived that

$$s = \frac{nvt}{nv + t} \,.$$

Differentiating with respect to n one gets

$$\frac{ds}{dn} = \frac{vt^2}{(nv + t)^2} \,.$$

Since in all scenarios of interest $nv >> t$,

$$\frac{ds}{dn} \approx \frac{t^2}{n^2 v} \,.$$

The change in s due to a small change Δn in n is given by

$$\Delta s = \frac{t^2}{nv}\frac{\Delta n}{n} \,. \qquad \textit{Equation 136}$$

The change Δs in s can be interpreted as s-t, and using Equation 132

$$b_c = \Delta s \frac{p}{s} = \frac{t^2}{nv}\frac{p}{s}\frac{\Delta n}{n} \,.$$

The variable s can be approximated by t in the denominator, to get

$$b_c \approx \frac{t}{nv} p \frac{\Delta n}{n},$$ *Equation 137*

and the corresponding limitation on the number of lines is

$$N_c = \frac{nv}{t} \frac{n}{\Delta n} = R_p \frac{v}{t} \frac{n}{\Delta n}.$$

The change Δn in the index of refraction over the visible spectrum in all polymers used for lenticular sheets is much less than 1. Therefore, the spot size corresponding to this aberration is much smaller compared to the other two, and can be neglected.

11.6 Mapping of lenticule optical data to a sheet

The optical properties of a lenticule in the viewing angle plane were derived in paragraph 11.4.6. In practice, one would like to find the distribution of the optical resolution and distortion over a given picture plane and for a given viewing distance. It is possible to associate vertical and horizontal viewing angles to any point ***P*** on the picture, as shown in Figure 177.

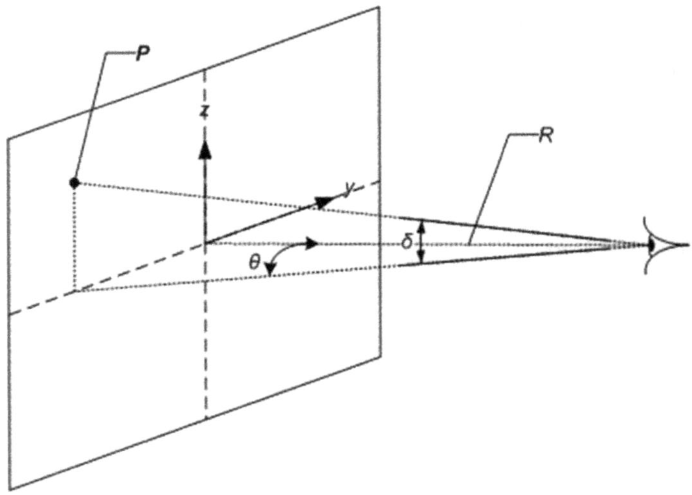

Figure 177: The transformation from angular to Cartesian coordinates.

Let the picture be at the *y-z* plane and the observation point be on the *x*-axis at a distance *R*. The task is to compute the horizontal and the vertical angles θ and δ for a given point ***P*** on the picture plane. The following relations follow from elementary trigonometry:

$$\theta = \tan^{-1}\left(\frac{P_y}{R}\right)$$

and

$$\delta = \tan^{-1}\left(\frac{P_z \cos\theta}{R}\right)$$

These relations can be used to map the picture aperture rigorously to the angles plane. The mapping of the angles plane to the Cartesian is not rectilinear. However, the deviations between the rigorous and the rectilinear mappings are small, and a rectilinear mapping can be used to obtain approximate estimates for resolution and distortion.

To estimate the optical resolution and distortion distributions directly from figures like Figure 171 through Figure 174, one may proceed as follows:

1. Calculate the horizontal and vertical angular extensions of the picture,
2. Form a rectangle with these angular sizes and superimpose it on the corresponding contour plot.

For example, suppose that the picture is a 40cmx30cm rectangle viewed from 100cm, as shown in Figure 178. The angular extensions of the picture will be 22.6°x17°.

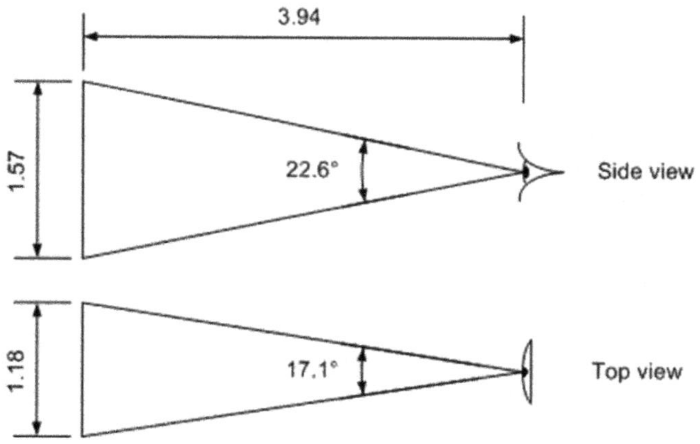

Figure 178: Angular extensions for the viewing scenario in the example.

Superimposing this rectangle on Figure 171 gives the result shown in Figure 179. One can conclude that for this scenario, and for a relative pitch of 0.5, that the optical resolution will be greater than 100, except for the extreme corners. The picture is also well within the acceptance angle interval so that viewing zone distortions may be disregarded.

The rectangle representing the picture is shown centered with respect to the plot, implying that the observation point is centered with respect to the picture. If this is not the case, one can shift the rectangle center to the angles defined by the picture center with respect to the observation point. In Figure 180 we show the same picture and viewing distance as in Figure 179, except that the center is shifted to a horizontal angle of 10°. In this case, the right edge of the picture becomes very close to the acceptance angle interval, and the optical resolution in its vicinity drops to approximately 40. This example demonstrates the advantage of viewing lenticular pictures from the center, both horizontally and vertically.

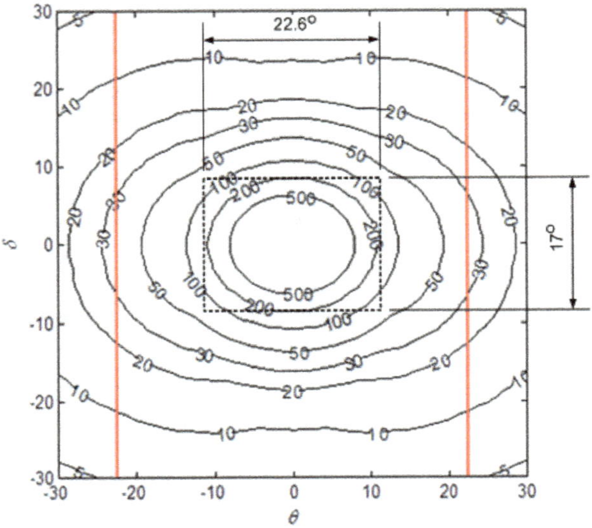

Figure 179: The optical resolution distribution for a 40cmx30cm picture centrally viewed from 100cm (the picture is represented by the dotted rectangle).

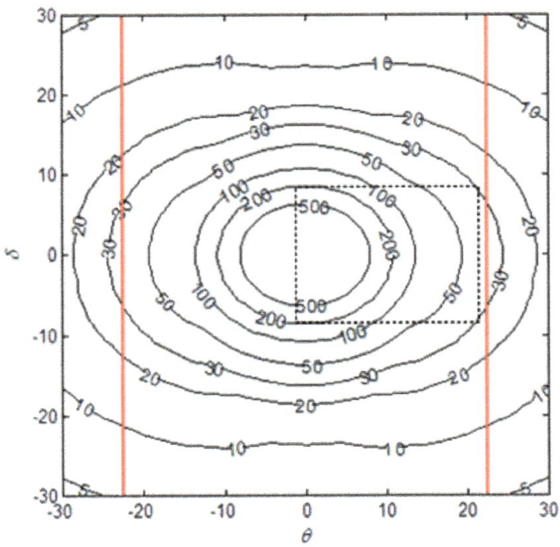

Figure 180: Same as Figure 179, but with a center horizontal viewing angle of 10°.

11.7 Discussion

In the vicinity of the picture center for central viewing (zero horizontal and vertical angles) the optical aberrations are very small. In this region, the optical resolution will be limited by the additional mechanisms considered above. From all three additional mechanisms that were considered, the dominant limitation on the number of lines that can be resolved comes from the finite pupil mechanism, and the corresponding limitation is given by Equation 135.

Let us consider a sizeable 3D picture with relative pitch of 0.5 viewed from 2000mm distance, and let the observer's pupil diameter be 5mm. This gives a limit of 200 resolvable lines. From Figure 171 it can be concluded that the additional mechanisms are dominant in the horizontal viewing angle interval of [-10° 10°]. For horizontal viewing angles larger than 10°, the contribution of the geometrical aberrations starts to take over.

The distortion becomes significant when its value becomes comparable to the printed sub-cell width of the interlaced picture. At a zero vertical angle, the relative distortions reach values of 0.002 and 0.01 at the edges of the acceptance angle intervals for relative pitch values of 0.5 and 1 respectively. Such distortions can become significant in pictures with sequences of more than 100 images. However, for a vertical angle of 30°, the relative distortions (also at the edges of the acceptance angle intervals) reach values of 0.01 and 0.04 for relative pitch values of 0.5 and 1 respectively. A relative distortion value of 0.04 may affect pictures with a sequence of 25 images or higher.

It should be noted that the other mechanisms discussed in 11.5 also contribute to the distortion. This effect was not evaluated.

It is possible to reduce the loss of resolution due to the geometric aberrations by using a non-cylindrical refractive surface. A proposal to use elliptic surfaces was put forward (Johnson & Jacobsen, 2005), and a significant improvement of resolution was demonstrated in the imaging plane.

Appendix: Main quantities and definitions

The main quantities used in the book are denoted with the same symbols in all chapters, although the use of symbols is not necessarily unique.

Table 29: Main quantities, definitions and symbols

Name	Symbol	Physical unit	Definition
Lenticule thickness	t	length	Figure 3
Refraction index	n	-	-
Relative pitch	R_p	-	Equation 6
Viewing angle	α	-	Equation 7
Acceptance angle	β	-	Equation 8
Lenticule pitch	p	length	Figure 3
Axial grid pitch	p_a	length	Paragraph 2.5
Optical grid pitch	p_o	length	Paragraph 2.5
Printed pitch	p_p	length	Paragraph 4.1
Limit of continuous imaging	h'	length	Equation 56
Limit of resolved imaging	h^*	length	Equation 60
Total parallax	P	length or pixels (depending on context)	Paragraph 6.3
Incremental parallax	P_s	length or pixels (depending on context)	Paragraph 6.5

Index

U

V

Bibliography

Berthier, A. (1896). Images stereoscopiques de grand format. *Cosmos*, 205-210, 227-233.

Byoungho, L., Soon-gi, P., Keehoon, H., & Jisoo, H. (2016). *Design and Implementation of Autostereoscopic Displays.* Bellingham: SPIE.

Castle, A. (2015). *How to fix your ingram spark pdf.* Retrieved December 21, 2017, from Self Publishing Review: http://www.selfpublishingreview.com/2016/03/how-to-fix-your-ingram-spark-pdf-with-free-pdf-fix-download/

Cathode Ray Tube. (2017). Retrieved November 15, 2017, from Wikipedia: https://en.wikipedia.org/wiki/Cathode_ray_tube

Christiansson, H., & Blom, E. (2012, November). Retrieved from Innventia: www.innventia.com

Cosima. (2017). Retrieved August 10, 2017, from Cosima: http://www.cosima-3d.de/

Dean, C. (2015). Photographing for 3D Lenticular. *Stereoscopy*(103), 1-3.

Depth map. (2017). Retrieved August 10, 2017, from Wikipedia: https://en.wikipedia.org/wiki/Depth_map

Flat panel displays. (2017). Retrieved November 15, 2017, from Wikipedia: https://en.wikipedia.org/wiki/Flat_panel_display

Foley, J. D., Andries, v., Freiner, S. K., & Hughes, J. F. (1990). *Computer Graphics.* Reading, Massachusetts: Addison Wesley.

Gardner, R. (2016). *Inkjet resolution.* Retrieved December 10, 2016, from Rags Int. Inc.: http://www.rags-int-inc.com

Halftone. (2017). Retrieved August 10, 2017, from Wikipedia: https://en.wikipedia.org/wiki/Halftone

Hartley, R., & Zisserman, A. (2003). *Multiple View Geometry in Computer Vision* (2nd ed.). Cambridge: Cambridge University Press.

Hess, W. (1915). *US Patent No. 1,128,979.*

Ives, F. E. (1902). *Journal of the Franklin Institute*, 51-52.

Ives, F. E. (1903). *US Patent No. 725,567.*

Jenkins, F. A., & White, H. E. (1957). *Fundamentals of Optics.* New York: McGraw-Hill.

Johnson, R. B., & Jacobsen, G. A. (2005). Advances in lenticular lens arrays for visual display. *Proc. SPIE. 5874.* San Diego: SPIE.

Kanolt, C. W. (1918). *US Patent No. 1,260,682.*

Knigin, S. (2016). Retrieved March 10, 2016, from 3dmix: http://www.3dmix.com/rus/articles/knigin2/

Lee, D. L., & Winslow, A. T. (1993). Performance of Three Image-Quality Metrics in Ink-Jet Printing of Plain Papers. *Journal of Electronic Imaging, 2*(3), 174-184.

Leuder, E. (2012). *3D Displays.* Chichester: John Wiley and Sons.

Lippman, M. G. (1908). Photographie: Epreuves reversible, Photographies integrales. *Comptes Rendus de l'Academie des Sciences, 146*(9), 446-451.

List of refractive indices. (2015). Retrieved December 2, 2015, from https://en.wikipedia.org/wiki/List_of_refractive_indices

Ma, Y., Soatto, S., Kosecka, J., & Sastry, S. S. (2004). *An Invitation to 3-D Vision.* New York: Springer-Verlag.

Mathworks. (2017). Retrieved November 5, 2017, from Mathworks: https://www.mathworks.com/

Mendiburu, B. (2009). *3D Movie Making.* Amsterdam: Focal Press (Elsevier).

Okoshi, T. (1976). *Three-Dimensional Imaging Techniques.* New York: Academic Press.

OpenCV. (2017). Retrieved November 5, 2017, from OpenCV: https://opencv.org/

Parallax barrier. (2017). Retrieved August 17, 2017, from Wikipedia: https://en.wikipedia.org/wiki/Parallax_barrier

Pratt, W. K. (1991). *Digital Image Processing.* New York: Jown Wiley and Sons.

Pseudoscope. (2017). Retrieved November 9, 2017, from Wikipedia: https://en.wikipedia.org/wiki/Pseudoscope

Refractive Index Info. (2015). Retrieved December 2, 2015, from http://refractiveindex.info/?shelf=organic&book=polycarbonate&page=Sultanova

Scientific Polymer. (2015, December 2). Retrieved from http://scientificpolymer.com/technical-library/refractive-index-of-polymers-by-index/

Shibata, T., Kim, J., Banks, S. M., & Hoffman, D. M. (2011). Ths zone of comfort: Predicting visual discomfort with stereo displays. *Journal of Vision, 11*(8).

Stereo Photo Maker. (2017, August 10). Retrieved from Stereo Japan: http://stereo.jpn.org/eng/stphmkr/

Stereopsis. (2017, August 17). Retrieved from Wikipedia: https://en.wikipedia.org/wiki/Stereopsis

Streefland, W. (2006, May). Minimum Printed Dot Size as a Print Capability Measurement. *International Paper Board Industry (UK)*, pp. 70-75.

The treachary of images. (1929). Retrieved December 12, 2017, from Rene Magritte: https://www.renemagritte.org/the-treachery-of-images.jsp

van Berkel, C. (1999). Image Preparation for 3D-LCD. *Stereoscopic Displays and Applications, SPIE Vol. 3639*, p. 84 to 91. San Jose, California.

van Berkel, C., & Clarke, J. A. (1997). Characterisation and Optimisation of 3D-LCD Module Design. *Stereoscopic Displays and Virtual Reality Systems IV, SPIE Vol. 3012*, pp. 179-186.

Vance, R. (2000). The Lenticular Legacy of Harvey Prever. *Stereo World, 27*(1), 1 to 6.

Vector Graphics. (2017). Retrieved July 10, 2017, from https://en.wikipedia.org/wiki/Vector_graphics.

Weissman, Y. (2012). Geometrical Principles of Orthostereoscopic Imagery. *Stereoscopy*(1), 27-33.

Wolberg, G. (1990). *Digital Image Warping.* Los Alamitos: IEEE Computer Society Press.

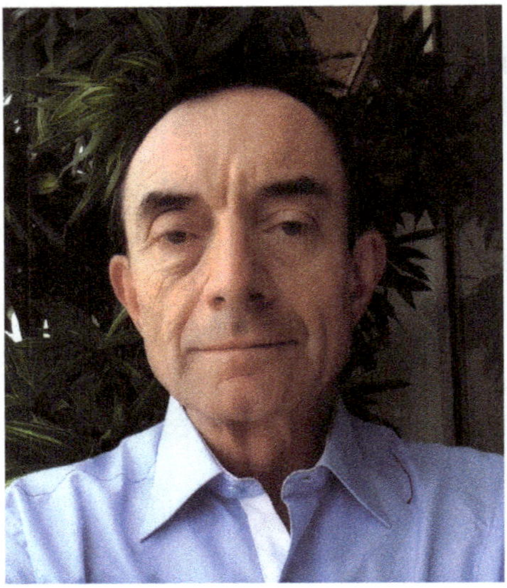

Dr. Yitzhak Weissman received his Ph.D. degree from Tel Aviv University in 1981. Since then, his career interleaved scientific and applied research.

During his academic activity, Dr. Weissman published over 30 scientific papers and two books: "Optical Network Theory" in 1991, and the present book in 2018. The subjects of his publications include solid-state physics, quantum mechanics, optical fibers, infra-red technology, optical engineering, and heat transfer. Dr. Weissman spent two sabbatical periods in the US: in the UC Berkeley Chemistry Department in 1981 as a post-doctoral fellow, and in the University of Utah Electrical Engineering Department in 1991 as a Visiting Professor.

Dr. Weissman's passion for innovation and entrepreneurship has led him to participate in numerous start-up projects, which included the development of fiber-optic sensors, infra-red spectroscopy instrumentation, simulator displays, optically efficient polarizers for projection 3D displays, diamond color grading, and lenticular print.

Dr. Weissman's involvement in 3D visualization started with stereo photography. His lenticular printing service of stereo photographs became very popular among stereo enthusiasts around the world. The work on lenticular stereo pictures led him to the study of 3D lenticular imaging, which is documented in the present book. The work on this subject shifted Dr. Weissman's attention to 3D lenticular photography, and to a branch of 3D digital art called "Virtual Sculpture." Nowadays, he is an active artist in these fields.

Dr. Weissman was born in Poland in 1949. In 1957 he immigrated with his family to Israel. He is married, has two children, and lives in Herzliya, Israel.

This page left intentionally blank

This page left intentionally blank

This page left intentionally blank

This page left intentionally blank

This page left intentionally blank

This page left intentionally blank

This page left intentionally blank

This page left intentionally blank

www.ingramcontent.com/pod-product-compliance
Lightning Source LLC
Chambersburg PA
CBHW071222230926
51849CB00001B/8
9789655724035